Managing

Managing Airline Networks: Design, Integration and Innovative Technologies is a fully comprehensive description of state-of-the-art network management practices at airlines. Designed as a compendium on current practices and future trends in the field, the book offers an instructive guide through the complex world of non-linear production systems.

Written by a renowned consultant and aviation expert, the book discusses the impact of network management on airline resource planning and performance, and examines the interplay between network management and adjacent functions. The book includes a practical case study and is enriched with academic perspectives. Discussing upcoming trends in the sector, the book provides an outlook on advanced technologies that may play a role in next-generation network management.

Features include:

- a description of basic network types, performance indicators for profitable networks, efficient processes and success factors for network management, and common optimisation models and tools;
- descriptive overviews, supported by practical examples, and leading to a deep-dive case study;
- a section on trends in network management, outlining new demand forecasting models, 'big data' applications, machine learning and AI use cases, and alternative optimisation models for airlines.

Managing Airline Networks: Design, Integration and Innovative Technologies is designed as a comprehensive compendium and is essential reading for both aviation practitioners and students of airline management.

Markus Franke is owner of Franke Aviation & Transportation Consulting (FATC) and Partner at M2P Consulting in Frankfurt. He has a track record of 29 years in aviation and 25 years in consulting, focusing on strategy and digitalisation. He is a lecturer at the International University of Bad Honnef (IUBH) and has published various articles and papers.

Managing Aviation Operations
Series Editor: Peter J. Bruce
Associate Editor: John M. C. King

The purpose of this series is to provide a comprehensive set of materials dealing with the key components of airline operations. To date, this innovative approach has not been evident among aviation topics and certainly not applied to operational areas of airlines. While more recent works have begun, in brief, to consider the various characteristics of operational areas, the Managing Airline Operations for Aviation Professionals series will expand coverage with far greater breadth and depth of content.

Airlines are devoid of specific topic knowledge in ready-made, easy-to-read, creditable resources. Tapping into industry expertise to drive a range of key niche products will resource the industry in a way not yet seen in this domain. Therefore, the objective is to deliver a collection of specialised, internationally sourced and expertly written books to serve as readily accessible guides and references primarily for professionals within the industry. The focus of the series editors will be to ensure product quality, user readability and appeal, and transparent consistency across the range.

Airline Management Finance
The Essentials
Victor Hughes

Airline Operations Control
Peter J. Bruce and Chris Mulholland

Airline Governance
The Right Direction
Victor Hughes

Managing Airline Networks
Design, Integration and Innovative Technologies
Markus Franke

For more information about this series, please visit: www.routledge.com/Aviation-Fundamentals/book-series/MAO

Managing Airline Networks

Design, Integration and Innovative Technologies

Markus Franke

Routledge
Taylor & Francis Group
LONDON AND NEW YORK

First published 2021
by Routledge
2 Park Square, Milton Park, Abingdon, Oxon OX14 4RN

and by Routledge
52 Vanderbilt Avenue, New York, NY 10017

Routledge is an imprint of the Taylor & Francis Group, an informa business

British Library Cataloguing-in-Publication Data
A catalogue record for this book is available from the British Library

Library of Congress Cataloging-in-Publication Data
A catalog record has been requested for this book

ISBN: 978-0-367-33238-9 (hbk)
ISBN: 978-0-367-33237-2 (pbk)
ISBN: 978-0-429-31879-5 (ebk)

Typeset in Bembo
by Newgen Publishing UK

Contents

Figures

Tables

Preface

Aims of this book

This book is a compendium for aviation practitioners, enriched with some academic perspectives. Furthermore, it provides an outlook on potential links between network management and its adjacent commercial functions (such as fleet management, hub optimisation and revenue management), as well as upcoming technological or procedural innovations. With this structure, the book offers a comprehensive overview of state-of-the-art network management paradigms and tools, and a rare combination of both practical and strategic concepts for airline professionals. The aim of this book is to empower these practitioners to efficiently manage airline operations and commercial functions by unleashing the full potential of advanced network management.

This book provides an overview of the main drivers relevant for sound network design and examines the performance indicators suited to measuring successful design and operations. Furthermore, it illustrates selected correlations between major drivers, and introduces common methodologies to solve the inherent trade-offs.

To round this compendium up, there are sections discussing the potential integration of network management with neighbouring functions, as well as providing an outlook on upcoming new technologies possibly relevant for network management.

Structure of the book

This book contains three main parts.

In Part I, a fundamental view on network management is given, including its definition, its instruments and its value for airlines, as well as a morphology of common network types. Subsequently, this part provides an introduction to state-of-the-art network management practices along the following aspects:

- measurement of successful network design and management;
- common network management processes;

- optimisation models and planning tools for networks; and
- external and internal frameworks for network management, such as regulation, fleet structure and hub concepts.

To deepen the understanding of the significance of network management, its impact on commercial success and operational efficiency is discussed. Part I is rounded off by a simplistic, but realistic case study on the choices and effects of network management in a competitive environment.

Part II describes the interfaces and interactions between network management and adjacent functions, such as fleet, hub or crew planning. The underlying assumption is that, with the exponential growth of computational power, a trend towards more comprehensive planning models may emerge, furthering a closer functional integration between network management and some of these functions. This part evaluates the probability that the respective integration may occur, discusses the prerequisites and deduces potential benefits from a closer integration.

Part III provides an outlook on trends in network management, as well as on the potential role of advanced technologies such as 'big data' or machine learning. This part is concluded by a perspective on selected, advanced mathematical theories that may have an impact at some point in time on the speed and efficiency of network optimisation models.

Acknowledgements

I would like to express my most cordial gratitude to the editors, Peter J. Bruce and John M. C. King, who supported me greatly in accomplishing this book. Without their kind encouragement and competent advice, the book would not exist the way it does.

Glossary and list of abbreviations

A/C	aircraft
ACMI	aircraft, crew, maintenance, insurance
AI	artificial intelligence (also referred to as deep learning)
AOC	air operator certificate
AOG	aircraft on ground
ASK	available seat kilometres
ASM	available seat miles (cf. ASK)
BESLF	break-even seat load factor
Bh	block hour (i.e., operating time of aircraft)
C	continental (short-haul)
CapEx	capital expenditure (cf. OpEx)
CASK	cost per available seat kilometre
CASM	cost per available seat mile (cf. CASK)
CAMO	Continuing Airworthiness Management Organisation
DOC	direct operating cost
EASA	European (Union) Aviation Safety Agency
EBIT	earnings before interest and taxes (operating profit)
EMSR	expected marginal seat revenue
F-RAN	fog computing-based radio access network (a technology for mobile telecommunication networks)
IATA	International Air Transport Association
IC	intercontinental (long-haul)
ICAO	International Civil Aviation Organization
ICN	information-centric networking (a paradigm for mobile telecommunication networks)
IP	integer (linear) programming (a mathematical optimisation method; cf. LP)
KPI	key performance indicator
LCC	low-cost carrier
LP	linear programming (a mathematical optimisation method)
MaxCT	maximum connecting time (in a hub)
MIDT	Market Information Data Tapes (a database within Sabre's AirVision service)

MILP	mixed integer linear programming (a mathematical optimisation method; cf. LP)
MIT	Massachusetts Institute of Technology (Cambridge, MA, United States)
MCT	minimum connecting time (in a hub)
MDP	Markov decision process (a mathematical reference model)
MTT	minimum turnaround time
MRO	maintenance, repair, overhaul (of aircraft)
NDC	New Distribution Capability (IATA data transmission standard)
NLP	natural language processing (a machine learning method)
NoSQL	not only structured query language (non-relational database language; cf. SQL)
NTP	normal tariff price (usually business class tariff; cf. STP)
O&D	origin and destination (starting and end point of a traveller's itinerary – airport-to-airport view)
OAL	other airlines
OCC	operations control centre
OEM	original equipment manufacturer (e.g., of aircraft or jet engines)
OLAP (cube)	online analytical processing (a multidimensional structure of data warehouses)
OpEx	operational expenditure (cf. CapEx)
P2P	point-to-point (network)
pax	passengers
RPK	revenue passenger kilometres
RPM	revenue passenger miles (cf. RPK)
RASK	revenue per available seat kilometre
RRPK	revenue per revenue passenger kilometre
SKO	seat kilometres offered (cf. ASK)
SLF	seat load factor
SQL	structured query language (a relational database language/protocol)
STP	special tariff price (usually economy tariff; cf. NTP)
true O&D	origin and destination (starting and end point of a traveller's itinerary – door-to-door view; cf. O&D)
WEF	World Economic Forum (a non-profit organisation from Geneva, Switzerland)
WHU	Wissenschaftliche Hochschule für Unternehmensführung (originally); today: Otto Beisheim School of Management (Vallendar/Koblenz, Germany)
WTI	West Texas Intermediate (crude oil grade)

Part I

State-of-the-art network management at airlines

The first part of this book draws a picture of current network management practices as they are executed today at leading airlines. The description includes chapters on network types, success factors and profit calculation schemes, related processes (from long-term to short-term) and optimisation models (including dedicated tools). Furthermore, the impact of network decisions on the commercial results and operational stability of an airline is discussed, and a simplified case study on typical network planning scenarios is provided.

1 Introduction

1.1 Definition of networks

In technical terms, the term 'network' refers to a physical (tangible) or virtual (intangible) structure consisting of 'nodes' and 'edges'. Examples of purely physical networks are computer LAN structures, utility networks (for power supply), railway systems and traditional landline telecommunication infrastructure; purely intangible networks are, for instance, private and professional relationships.

There are also hybrid networks, such as mobile telecommunication and air transportation systems. In both cases, the 'node' is a physical infrastructure (a transmitting station or a mobile phone, an airport), while the 'edge' represents a virtual connection between two nodes.

This book deals with air transportation networks. Networks are a core production factor of airlines, accounting for the majority of all revenues and costs generated in the course of airborne transportation.[1] There are various assets and resources needed to operate the network, such as aircraft, crews, airports and airspace for airlines. There are two characteristic aspects of a transportation network:

- the production model within the network is *non-linear*; and
- the involved assets and resources are highly *interdependent*.

Thus, a meaningful definition of a transportation network could be the following:

> *The non-linear interplay of interdependent assets and resources to move passengers or cargo from various origins to various destinations.*[2]

Linear production models, such as car manufacturing, can be broken down into single, sequential production processes or value chains: raw materials are transformed step by step into intermediate products, and, ultimately, into finished products. At a certain stage of development, the amount of raw materials cannot be changed, nor can the sequence of the value-adding steps. On a higher level, there are in any case network-like structures (e.g., various production plants in different geographies), providing the management with different options for where and when to allocate resources to generate the desired output. At the 'atomic' level, however, there are no choices once the production technology and process have been determined.

1.2 Non-linear production models

This is clearly different in *non-linear* production models. The smallest conceivable transportation system would be the connection between two points (for airlines: two airports, one route). Even in this embryonic 'network', the management would have a variety of non-trivial options as to how to allocate resources:

- the number of services offered per day/week/month ('frequency');
- the timing of services offered;
- the size/seat capacity of aircraft;
- the seat configuration within aircraft (cabin classes); and
- the service components offered on board and on the ground.

These options are, on the one hand, positive, since they allow adjustments to concrete market environments and customer needs. On the other hand, this variety comes at a price, since it creates planning complexity.

Another complication is caused by the permanent real-time effect of competition on the production process of an airline. While a car manufacturer will realise the effect of competing products no earlier than its cars are finished and offered in the market (not *during* the production process), transportation companies – and particularly airlines – experience a permanent interaction between their own production process and those of other airlines. For instance, if a competitor decides to depart 15 minutes earlier on a certain route, and deploys a different aircraft (smaller or larger), this may alter the production model of an airline on that route from initially efficient to inefficient – without any internal change.

Adding just one single node to the above-mentioned embryonic network of two nodes increases the inherent complexity, since it adds more options – not to speak about dozens or even hundreds of nodes as offered by major airlines. As a tendency, the number of dimensions to be tackled in a production plan shows at least parabolic growth with the number of nodes, since the number of potential routes r in a network can be calculated as $r = n \cdot (n–1) / 2$ (with n being the number of nodes). In practice, the correlation can even be exponential, since there are many more influencing parameters than just the number of nodes.

Network-like structures exist in many industries, as well as in private domains – including the current mega-trend of social media networks. What are the characteristics of airline networks? There are many definitions, but the constituting element of all transportation structures, including airline networks, is the interdependence between core resources. Consequently, these resources cannot be planned separately, but in conjunction with related resources, which causes complexity.[3]

The complexity is also impacted by the business model chosen by a certain airline, however. For *low-cost carriers* (LCCs), for instance, the parabolic correlation between number of nodes and number of routes is fully valid, since this model relies on simple point-to-point (P2P) connections. The *interdependence* of assets and resources is limited, though, since low-cost carriers try to avoid situations in which one aircraft rotation depends on another one. In total, the complexity inherent in the network of low-cost carriers is lower than that to be mastered by *hub and spoke airlines*.

In the business model of the latter, nodes are not all equal: one central airport, the so-called *hub*, is specifically designed to cumulate demand and feed all other nodes (referred to as 'spoke airports' or 'outstations'). Consequently, hub and spoke carriers are able to cover *n* nodes with only *n–1* routes, which establishes a much simpler (i.e., linear) correlation than in the case of the point-to-point network. On the other hand, the reduced complexity through a lower number of routes is replaced by the increased complexity within the hub, whereby the resources are highly *interdependent* (the impact of the chosen business model will be detailed in Chapter 2).

1.3 Definition of network management

In essence, *non-linear* production models entail significant complexity, driven by the fast growth of dimensions, and the interdependence of assets and resources. This leads to a highly complex mathematical and operational optimisation problem. Since there is no fully integrated mathematical solution for this problem so far, airlines apply a sequence of stepwise-designed paradigms and planning algorithms to ensure appropriate fleet structures, attractive route patterns, competitive service levels and efficient airport operations.[4] In addition, network scenarios will be compared and refined until the network contribution appears to be optimal.

Since there is not a single ideal network design, the scenario evaluation needs to take into account the specifics of each airline, such as the regulatory framework, home market and shareholder structure. Potential conflicts between short-term and long-term objectives have to be carefully managed. Shareholders and managers will definitely aim for sustainable profits, but also for satisfied travellers, attractive jobs and ecological sustainability. Considerable regulation, as well as structural overcapacity in the markets, reduces the leeway in this exercise and imposes an additional burden on airline managers.

Concluding this chapter, the following definition of network management addresses all relevant aspects:

> *The allocation of scarce, interdependent resources to a complex transportation process under uncertainty.*[5]

The following chapters will build on this definition. In particular, the three concepts '*interdependence*', '*complexity*' and '*uncertainty*' are fundamental and indispensable for an understanding of airline network management.

1.4 Value of network management for airlines

Before airlines are able to offer any kind of air transport services to their clients, they need to foresee upcoming demand for airborne mobility, and define a certain number of flights on specific routes they intend to operate. As stated earlier in this chapter, one of the constituting elements of airline networks is the interdependence between core resources, causing significant complexity.

An aircraft's daily itinerary ('rotation') may, for instance, differ from the respective rotation of a crew licensed to operate this aircraft, on account of different operational limitations, such as the maximum duty times of crew members. Nonetheless, it needs to be ensured that an appropriate crew is available on each segment of the aircraft's rotation ('leg'). This kind of interdependence is valid for all other resources as well, including, for example, ground handling staff, or aircraft maintenance capabilities.

The larger an airline's network, the more complex the calculation will be. Possible combinations of resources will show exponential growth with the size of the network. The most complex case occurs if an airline decides to operate a 'hub and spoke' network, cumulating demand flows at one or more central airports. Incoming flights to some extent 'feed' outgoing flights in this concept, connecting clients to more worldwide destinations by transferring them in this 'hub'. Through the 'feeding' approach, hub and spoke networks establish an even higher level of resource link, and, thus, planning complexity (for details, see Chapter 2).

In an ideal world, the described complexity would be handled by a mathematical model delivering the perfect combination of all involved resources at each point in time. There is no such model, though, and, even if one were available, current information technology (IT) is not advanced enough to compute the required solutions at sufficient speed.[6] The latter restrictions may change, but the basic limitation will persist: a single optimum solution for a network problem can be calculated in theory but should not be targeted in practice.

Airline networks are operated under specific conditions (e.g., regulatory framework, labour laws, home market, or shareholders), pursuing a specific set of objectives. The maximum profit may not be achievable with a given fleet, or a scenario that appears favourable in the short run may collide with strategic goals. In the absence of a fully integrated network calculation model, airline managers usually deploy partial or sequential solutions, slicing the problem into manageable portions. Specialised optimisation algorithms exist, for example, for fleet planning, network design, rotational planning, crew planning and hub operations. These solutions still imply massive complexities, but are easier to handle in a stepwise approach than in one unified model (for details, see Chapter 4, sections 4.3 and 4.4).

As already stated earlier in this chapter, the particular value of network design and planning lies in the fact that networks account for the vast majority of revenues and costs an airline may have, irrespective of the specific network model or calculation approach. Some airline functions assume a cost perspective, while others have a revenue view on air transport; network planning is the only function that combines both perspectives. Network profit calculation schemes are thus the core of an airline's profitability calculation (for details, see Chapter 3, section 3.3).

Notes

1 Franke, M., 'Network design strategies', in Bruce, P. J., Gao, Y., and King, J. M. C. (eds.), *Airline Operations: A Practical Guide* (44–61), Abingdon: Routledge, 2018.
2 Franke, M., lecture, 'Strategic network management', IUBH, academic year 2019/20.
3 Franke, 'Network design strategies'.
4 Franke, 'Network design strategies'.
5 Franke, 'Strategic network management'.
6 Franke, 'Network design strategies'.

2 Planning paradigms for different types of networks

As already stated in Chapter 1, the optimum design and operations of an airline network depend strongly on the business model chosen by this airline, as well as on the type of network operated. There are three major network types found in practice – with a variety of sub-types and hybrid forms:

- hub and spoke networks (passenger and air cargo);
- point-to-point networks (passenger and air cargo); and
- gateway networks (express cargo).

2.1 Hub and spoke networks

Airlines operating hub and spoke networks boost the number of economically viable destinations by accumulating demand in one or more central airports, transferring passengers to and from flights that would otherwise have too little demand for a good service quality (i.e., a minimum frequency of flights per day/ per week). This approach is particularly relevant for long-haul intercontinental destinations, which very often could not be served point to point in a commercially viable way. The hub and spoke network type is the preferred model of traditional full-service carriers, and has the benefit of covering almost every possible revenue stream available in the markets served by this airline. Furthermore, the number of services needed to cover a certain number of destinations is as low as it can be: n destinations (besides the hub itself) can be connected by n flights using a hub.[1]

The benefits of this concept come at a cost, however. Hub and spoke networks are more complex than point-to-point networks, since the interdependence of resources is significantly stronger. For example, the envisaged feeder effect forces outbound flights to wait for a variety of inbound flights to collect all transfer passengers. This causes relatively poor productivity in terms of aircraft, plus operational risks: if one or more feeder flights are late, the outbound flight will – subject to operational discretion – automatically be delayed as well.

This 'domino' effect becomes even more powerful if a hub and spoke carrier structures its inbound and outbound hub flights in a special format called 'banks'. In this concept, flights are compressed into one or more 'peaks' to increase the probability that an inbound passenger can reach many outbound flights, a performance aspect of a hub referred to as 'connectivity'. The timing of the peaks observes criteria such as operational viability and passenger preferences. An operational driver, for instance, would be the availability of a slot at this airport at a certain time, which is by no means trivial at very busy mega-hubs. Furthermore, a matching slot must be available at the origin (for inbound flights) or destination airport (for outbound flights).

The term 'passenger preferences' usually refers to the needs of business travellers, since they are more time-sensitive than leisure travellers. For them, early morning and late evening flights are attractive, because this pattern maximises the daytime they can use at their destination to do business. In contrast, leisure travellers accept less attractive departure times such as during nighttime or around noon – rather unpopular for business travellers – because they hope for corresponding discounts. In fact, airlines reward these off-peak bookings, because it helps them to reduce the morning and evening peak load.

A peaked hub structure with clear, distinct banks basically ensures good hub performance in terms of connectivity and fleet productivity, provided that the chosen bank structure reflects the operational framework and scale of the respective hub well. On the other hand, operational peaks show a clear tendency to deteriorate asset and resource utilisation, and they increase the risk of operational instability. The respective trade-offs need to be managed efficiently to ensure an overall good result (for details, see Chapter 6, section 6.5).

2.2 Point-to-point networks

A point-to-point (P2P) network is the favourite production concept of low-cost carriers (LCCs). It is the purest way to form a network, though not the leanest one. Whereas hub and spoke networks are suited to connect n destinations with only n flights (via a hub), the P2P network requires $\frac{1}{2}\, n \cdot (n-1)$ flights. On the other hand, the interdependence of resources is much smaller than in the hub and spoke model, resulting in significantly lower complexity.

In the purest version of a P2P model, each aircraft commutes between just two destinations in a 'ping-pong-like' manner, minimising the link between different routes. Moreover, low-cost carriers in general try to avoid a hub-like structure, since every transfer of passengers induces waiting times between outbound flights and their feeder services – an effect that spoils aircraft productivity and is by no means 'low cost'. Finally, P2P networks are structurally more robust than hub and spoke networks, since they lack the inherent operational instability of hub concepts.

The boundaries between the above-mentioned business models have been blurred in the recent past, however, leading to both hybrid business models and network patterns. Hub and spoke carriers have always offered P2P services on selected 'trunk' routes with high P2P demand and business traveller share, such as London Heathrow (LHR) to New York's John F. Kennedy (JFK), Frankfurt (FRA) to Paris Charles de Gaulle (CDG) or Melbourne (MEL) to Sydney

(SYD). In the light of the continuing success of low-cost carriers, and declining yields (revenue per seat), full-service airlines felt forced to adopt more and more elements of the low-cost business model, or even to found low-cost branches suited to defend continental traffic shares. On the other hand, the rapid growth of low-cost carriers slowed down in the same time span, forcing them to tackle new sales opportunities, especially in the business travel segment. To attract this client segment, though, low-cost carriers need to raise their service level, and ultimately abandon certain beliefs of their lean operations model.

Consequently, the maturing aviation industry experiences a convergence of business models, reflected also in an assimilation of network designs. The latest evidence of this trend is the emergence of low-cost long-haul business models, introduced around 2015 by low-cost carriers such as Norwegian, but also by full-service carriers such as Lufthansa and Singapore Airlines (with their subsidiaries Eurowings and Scoot, respectively).[2] Long-haul services require at least some feeder traffic, so low-cost long-haul models need to be combined with low-cost hub operations.

The ongoing commoditisation trend, driven by both low-cost and hub and spoke carriers, has absorbed two other traditional airline business models: the charter business (mainly P2P) and regional operators (P2P and hub feed). Charter traffic has become a shrinking niche, since – especially short- and mid-haul – warm water destinations have been conquered by low-cost carriers. Regional carriers are in even worse shape, since they have mostly lost their commercial viability. Formerly high yields on selected secondary routes have eroded in a way that makes it almost impossible to cover the relatively high unit costs of regional jets or turboprop aircraft. Reflecting overall yield decline and cost pressure in the industry, mostly driven by the expansion of the low-cost model, the regional sub-network concept has almost disappeared, or been forced into tiny niches.

2.3 Gateway networks

The gateway network model is the typical operational model of global express delivery companies ('integrators') such as FedEx, UPS and DHL. Gateways connect regional or continental networks, ensuring a very high density for local pick-up and delivery with each other on a global level, through long-haul flights between the gateways. As opposed to passenger or air cargo carriers, integrators do not primarily sell free capacity but a certain quality level, which makes the regional high-density networks indispensable.

For example, DHL Express operates gateways in Wilmington (NC) for North America, as well as in East Midlands (UK) and Leipzig (Germany) for Europe. A parcel from any point in the United States to any point in Europe will be picked up by truck around noon, forwarded to Wilmington by large truck or aircraft in the evening, brought across the Atlantic to East Midlands or Leipzig with a wide-body aircraft during the night, then shipped to its final destination by aircraft or truck the next day.

2.4 Comparison of network types

Figure 2.1 summarises the characteristics of the three main network types, closely correlated with respective airline business models. In each of the three cases, one of the core functions of an airline has a dominant role: network design and planning for hub and spoke carriers (i.e., the hub and spoke model), lean operations for low-cost carriers (i.e., the point-to-point model) and commercial planning for integrators (i.e., the gateway model). In other words: the chosen business model determines the significance of the network planning function for an airline.

The correlation between business model and network type is strong, but not imperative. For example, large low-cost carriers may expand into a scale position such that their home base automatically assumes a hub-like role (e.g., Southwest Airlines in Dallas, Houston and Phoenix; referred to as 'self-hubbing of travellers'). Likewise, hub and spoke carrier may absolutely operate local point-to-point networks when connecting travellers via a hub would not be attractive enough against the direct services of competitors. Most airlines are by no means dogmatic about their preferred business model, but flexible at least in selected markets – as long as the extra benefit of the exception is not eaten up by the inconsistency caused by the exception.

Although low-cost carriers at least tended to be puristic about their business model in their early days (1990 to 2000), the past years since around 2005 have witnessed an ongoing convergence between the hub and spoke and the low cost business models. After the double dip of the global aviation industry

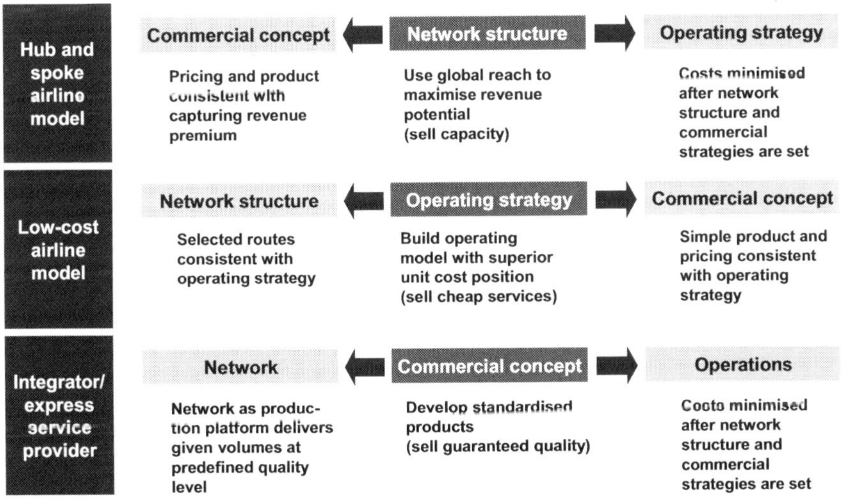

Figure 2.1 Characteristics of the three main network types[a]

[a] Franke, M., lecture, 'Strategic network management', IUBH, academic year 2019/20.

in 2001 and 2003, hub and spoke carriers were forced to copy elements of the low-cost model to survive the challenge of the rapidly growing low-cost carriers. Bringing down their cost level gave them the urgently required room for manoeuvre against the smart and cheap newcomers.

Vice versa, aggressive growth forces many low-cost carriers to adopt aspects of hub and spoke networks that they would have avoided by all means beforehand:

- Once the (point-to-point) network of a low-cost carrier has reached a certain size, travellers start to create their own connections in the home bases of the carrier ('self-hubbing'; see example of Southwest Airlines above). This behaviour entails a certain complexity and operational instability – the mere opposite of what a 'traditional' low-cost carrier would aim at. To minimise the complexity penalty, however, Southwest Airlines has shifted the connection risk to the self-hubbing traveller, and smartly asked for an extra baggage handling fee.
- Low-cost carriers initially focused on price-sensitive travellers such as heavy bargainers, backpackers, job commuters and ethnical travellers. After almost three decades (in the United States, even more than four decades) of massive growth with these preferred segments, low-cost carriers reached the ceiling on most routes, making further growth with the traditional customer segments very difficult. Consequently, major low-cost carriers at some point in time started offering service packages for business travellers as well, including features such as preferred boarding, increased seat pitch, extra baggage and refundable tickets. Some low-cost carriers even established loyalty programmes with additional benefits for frequent travellers – formerly an exclusive element of full-service business models that low-cost carriers would have never adopted in their early days because of the extra costs.
- Targeting business travellers intuitively leads to targeting corporate clients, as low-cost carriers have done; many of them have built up partnership programmes for large companies, including features such as special rates and kickbacks.
- Although low-cost carriers initially avoided partnership models – intensively operated by hub and spoke carriers – such as code sharing or alliances, growing competition has forced many of them in the meantime to go for partnerships and mergers. For example, Ryanair has intermittently collaborated with its Irish rival Aer Lingus, a hybrid carrier with a trans-Atlantic arm, to feed long-haul traffic.

In essence, the hub and spoke and the low-cost business model are still rivals, but they have learned from each other, mutually adopting successful features from their opponents. As a consequence, the continental service levels of both models have become remarkably similar. In the aftermath of the 2020 global COVID-19 crisis, it is a likely scenario that many airlines disappear from the market, or get acquired by other airlines. Likewise, airlines with different business models may enter collaborations that would have appeared unimaginable before the crisis. In the course of forced consolidation, low-cost carriers could, for instance, take over the continental feeder network of a hybrid type of alliance, while the hub and spoke partner focuses on intercontinental long-haul services.

2.5 Historic evolution of network management paradigms

For hub and spoke carriers, the complexity of their networks is a constituting element of both their business model and their network designing and planning approach. To better understand this perspective, a brief overview of the historic evolution of network management paradigms may be helpful.

The current network management practice described in this first section of the book is now some 25 years old. The last genuinely disruptive innovation was the move from a leg-based to an O&D-based perspective on transportation networks, early in the final decade of the past century. 'Leg-based' in this context means that every single flight segment ('leg') that a client may use on his itinerary is separately viewed and optimised by the respective airline, while 'O&D'-based network management takes the entire itinerary of a client (from origin to destination) into consideration.

For example, a client living close to Philadelphia (PA) may wish to travel to Milan in Italy, so his O&D (*airport to airport*) is Philadelphia to Milan, while his *true O&D* (*door to door*) is his small home town, close to Philadelphia, to the small home town of his Italian business partner, close to Milan. The traveller is interested in getting to his destination as seamlessly as possible – ideally, with a non-stop flight – but, unfortunately, there is no direct flight available between those cities ('city pair'). Thus, the traveller is forced to choose between offerings of different airlines with one stop or more, resulting in multi-segment or multi-leg flights. He could, for instance, book the following ticket (at the time of writing):

- flight from Philadelphia (PHL) to Chicago O'Hare (ORD), operated by United Airlines (UA);
- transfer in United Airline's hub; flight from Chicago O'Hare (ORD) to Frankfurt (FRA), operated by United Airlines;
- transfer in Lufthansa's hub; flight from Frankfurt (FRA) to Milan Malpensa (MXP) operated by Lufthansa (partner of UA).

Alternatively, he could book the following option:

- flight from Philadelphia (PHL) to London–Heathrow (LHR), operated by British Airways;
- transfer in British Airways' hub; flight from London–Heathrow (LHR) to Milan Malpensa (MXP), operated by British Airways.

There are many more options to cover the O&D demanded by the client with one, two or even more stops. Most passengers – especially business travellers – prefer the option with the smallest number of stops, since that saves time and is more convenient, and less subject to disruption. Nevertheless, a client could still decide to book a flight with more stops, for instance because he is a loyal client of the respective airline, enjoying frequent flyer benefits, or because he likes the transfer experience more in that hub, or because this flight is simply cheaper.

For our example, let us assume the client goes for the second option, booking British Airways' offering with two flight legs. There are many other passengers sharing the two flights with our example client; a few share the entire O&D Philadelphia to Milan with him, whereas most other passengers have only one flight with him in common. The latter clients may have their final destination in London, or go beyond London to a destination other than Milan, or come to London from an origin other than Philadelphia.

2.6 Historic approach: leg-based planning, pro-rating

Before the O&D-based approach was introduced, British Airways would have looked at all flight legs in its network separately (e.g., Philadelphia to London–Heathrow and London–Heathrow to Milan), trying to maximise the operating profit for all of them. The operating profit of a single flight can be defined by the revenues generated on this flight, minus the direct costs of this flight, minus some indirect costs assignable to this flight. Although the cost allocation seems to be straightforward – following the network profit calculation scheme introduced in Chapter 3, section 3.3 – the allocation of ticket revenue is unclear. If our client from the example above has booked a two-way Business Class ticket for US$4,000, British Airways could define that 50 per cent (= US$2,000) is allocated to either direction. But what share of the US$2,000 can be assigned to the first flight segment (PHL – LHR) and how much to the second leg (LHR – MXP)? A popular solution for this problem was the so-called *pro-rating approach*, indicating that the overall revenue was allocated *pro rata* in terms of *flight distance*.

The great circle distance between Philadelphia International Airport and London Heathrow is 5,700 km, while the distance between London Heathrow and Milan Malpensa is 940 km. Pro-rating leads to a revenue split of 86:14 per cent in relative terms, or US$1,717/US$283 in absolute terms, between the long-haul and the short-haul segments.

If we now assume that British Airways allocates direct and indirect cost amounting to US$1,100 for this incremental passenger to the long-haul flight, and US$400 to the short-haul-flight (73: 27 per cent in relative terms), we could come to the conclusion that lifting up this single passenger is profitable on the long-haul segment (profit = US$1,717 – US$1,100 = US$617), but unprofitable on the short-haul flight (US$283 – US$400 = –US$117). Unfortunately, this traveller would not have booked the entire trip with British Airways without the flight from London Heathrow to Milan, so this conclusion is misleading, if not wrong: ultimately, the incremental profit on the O&D level for lifting up this client is *positive* (US$2,000 – US$1,500 = US$500). Unlike the impression at first glance, the continental de-feeder flight does not, in fact, dilute the value of the long-haul flight, but makes it achievable!

One single case with an incrementally positive result would of course not justify sustaining a structurally loss-making feeder or de-feeder flight. This phenomenon occurs quite often on popular O&Ds, though, so the network effect described above for *one passenger* may easily be valid for the *entire O&D*: the de-feeder flight (or feeder flight, respectively) may look loss-making if the revenue pro-rating method is applied, but the long-haul and short-haul flights together create a (network) profit.

2.7 Downsides of the pro-rating method

The misleading results of the leg-based perspective can be easily explained: the pro-rating of revenues to flight segments based on flight distance is an intuitive, but overly simplistic method to calculate the profit of a single flight segment. This method is based on the assumption of a linear correlation between mileage and cost, which is particularly incorrect for long-haul flights. The direct costs of a flight are disproportionately dependent on the take-off and landing part of the flight, namely a considerable share of the fuel consumption, plus all airport fees, while the costs go down considerably during cruise. In this stage of the flight, only fuel consumption, crew costs and aircraft financing costs need to be taken into account. Thus, direct flight costs show a *degressive*, not linear, correlation with flight distance, the effect of which becomes more obvious the longer the flight distance. For instance, if a short-haul flight covering a distance of 1,000 km generates direct costs amounting to US$20,000, a long-haul flight with a distance of 7,000 km will not generate seven times the cost of the short-haul flight but, rather, say, three times – i.e., US$60,000 in this case.

A linear correlation between flight distance and direct flight costs is consequently a bad assumption, and splitting the O&D revenue accordingly – by pro-rating it – is an imprecise allocation method. Pro-rating revenues by distance systematically overrates the actual costs on long-haul flights, and thus allocates too much revenue – and vice versa for short-haul flights. In the example described above, the actual cost share of the long-haul flight PHL – LHR is 73 per cent, while pro-rating allocates 86 per cent of the ticket revenues to it. This systematic distortion makes the long-haul flight appear too positive and the short-haul flight too negative in terms of profit.

Before the O&D-based calculation approach was introduced at most airlines around 1995, network planners of course knew about the distortion described above. The leg-based approach with revenue pro-rating was intriguingly simple, but systematically wrong. Their solution was to introduce onward correction factors, limiting the impact of overrated/ underrated costs. In our above example, British Airways planners would have known that the LHR – MXP segment is a valuable de-feeder service for many profitable long-haul flights, adding a respective onward bonus. Likewise, they would have applied a penalty to the long-haul flight to ensure a more realistic profit. This method is suited to considerably improve the initial – often misleading – result of a leg-based planning exercise, but it is highly complex, since there are so many combinations between flight segments and O&Ds.

Ultimately, this solution requires a matrix with thousands of correction factors for all conceivable combinations between flight segments and O&Ds, which is a considerable effort. Furthermore, the factors must be regularly updated or recalibrated, to account for changes in the business environment. Applying correction factors to leg-based calculations is designed to alleviate, but by no means to cure, the fundamental problem of leg-based network management: costs occur on flight segments, while revenues occur on O&Ds, and it is never fully correct to bring them together at a leg level.

2.8 Current approach: O&D-based planning

The distortions caused by the leg-based planning approach described above create an increasing number of problems, especially for larger airlines. The more complex a network is, the more severe are the misallocations caused by the simplification of the leg-based approach. It became obvious that the non-linearity of the strongly interlinked production model of a hub and spoke carrier required a new approach. Consequently, scientists and airline network planners alike started to explore the concept of O&D-based planning. The remedy was simple on a conceptual level: if *revenues* cannot be properly split up to the *leg level*, *cost* must be aggregated to the *O&D level*. This solution created a variety of new problems, however.

- Although 'flight leg' and 'segment' – as the nucleus of every airline network – are physical categories, 'O&D' is a virtual concept, since it exists only in the head of a traveller. It is more difficult to plan resources on a level that is not a physical element of a production model.
- There are many more O&Ds than flight legs in the network of a sizeable airline, which makes calculations more complex. If an airline, for example, offers services to 100 destinations worldwide, it features no more than 100 flight legs as a hub and spoke carrier; the number of served O&Ds, however, can go up to 100^2 – i.e., 10,000.
- Consequently, O&D-based planning models are larger, and require more computational power and runtime.
- The major advantage of the O&D-based planning paradigm is that profit contributions or profits are correctly calculated on the O&D level. Unfortunately, this calculation does not provide any decision support for a single flight.

Planning a transportation network as a whole (i.e., taking all possible O&Ds into consideration) generates a complex resource allocation problem.

The network consists of a certain number of *nodes* (the destinations) and *edges* (potential routes between destinations). As stated earlier in this chapter, a hub and spoke network requires fewer routes than a point-to-point network to connect n nodes (n instead of $\frac{1}{2}\, n \cdot (n-1)$); the number of different options to move a passenger or shipment from one point (origin) to another point (destination) in the network is still enormous, however.

The number of options to allocate demand to capacity (aircraft seats) and capacity to routes very quickly exceeds the capability of a human planner to evaluate the different options and find the most efficient solution. For this kind of mathematical problem, the scientific domain known as operations research (OR) has been developed. The standard OR approach for transportation (network) problems is linear optimisation or linear programming (LP). This methodology is further explained in Chapter 5.

2.9 Computation of O&D-based planning

Luckily, the computational power to solve complex transportation problems has exponentially grown over the past 30 years. Thus, network scenarios can nowadays be calculated in minutes, whereas they would have required hours or days back in, say, 1995 – if they could have been solved *at all*. This progress is also key to tackling the challenge of calculating a single flight in a holistic O&D-based approach: as explained above, the O&D-based approach delivers accurate results, but only on an O&D or network level. It does *not* provide an explicit solution for the question as to whether an incremental passenger should be taken onboard a single flight, or whether this flight should be operated at all. A short-haul feeder or de-feeder flight may look unprofitable at first glance, but still have a positive contribution to the overall network – through high-value transfer passengers who pay higher-yielding fares for long-haul connections.

The only accurate solution is, in mathematical terms, a 'brute force' approach, but it works perfectly as long as enough calculation capacity and speed are available:

profit (passenger A) = profit (network with passenger A) – profit (network without passenger A)

profit (flight X) = profit (network with flight X) – profit (network without flight X)

The value of an incremental passenger or flight is calculated in an implicit way by comparing the profit of the entire network with and without that passenger or flight. The discrepancy in value must necessarily be the value of that passenger or flight if all other parameters stay unchanged ('*ceteris paribus*'). Admittedly, calculating two entire network scenarios just to determine the value of a single flight, or even passenger, sounds like cracking a nut with a sledgehammer. Thus, this accurate method will be applied rarely for a single flight, and almost never for a single passenger. Today's computer performance makes this approach available, though, and it may be good to know that simpler, less accurate calculation methods such as rules of thumb or heuristics can be verified or calibrated from time to time with a correct result.

In daily practice, the holistic approach is not very common – because of the complications mentioned above. Instead, the traditional solutions or simplified calculation methods described for the leg-based era are still being applied. This is particularly true for the calculation of a single passenger's value: during network design and scheduling, it is not really necessary to know the incremental value of a passenger. It is absolutely sufficient to assume a certain seat load factor and fare mix for all flights under consideration. The shorter the time till the departure of a flight, the more relevant a single passenger's contribution becomes. This question is shifted by most airlines from network management to revenue management, the short-term 'neighbour' of network management.

2.10 Comparison: O&D-based approach in revenue management

In revenue management, the simplifying assumption being applied is that, at very short notice, almost every cost item related to uplifting a single passenger is already fixed (i.e., a 'sunk cost'), and thus irrelevant for decision making. Consequently, every incremental passenger will be accepted if (a) he or covers at least the direct cost of his uplift (see network calculation scheme in Chapter 3, section 3.3) and (b) he does not displace any other incremental passenger ready to pay a higher fare. Network effects, especially those between low-value feeder flights and high-value long-haul flights, are being taken into account by correction factors or so-called onward construction rates.

In practice, the move from the historic leg-based approach to the more advanced O&D-based methodology was not performed in one big leap, but in different intermediary steps. Both network and revenue management operated in-between solutions, such as 'leg-based, O&D-oriented' planning. With growing computational power and the increasing sophistication of mathematical models, more and more simplified solutions could be skipped, until, ultimately, pure O&D-based planning models became available.

2.11 Network management at low-cost carriers and integrators

For business models *other than hub and spoke*, network design and planning are in any case an impactful function, but not as much a core activity as for hub and spoke carriers. For low-cost carriers, network design is, by definition, much less complex than for hub and spoke carriers, thanks to the absence of hubs and the dominance of P2P traffic. Instead, lean operations are of the essence, so that the business model of LCCs is dominated by operational management. Around 2010, however, LCCs started making compromises in terms of their business models, in the course of aiming at new client segments such as business travellers. Furthermore, once the networks of LCCs had reached a critical size, clients (for instance, those of Southwest Airlines) started creating their own connections ('self-hubbing') in the home bases of these airlines – an effect that network managers could not fully ignore.

Global gateway-based networks of integrators are mainly driven by the quality levels promised to the clients. Seamless capacity and speedy operations are of the essence, rather than high load factors or lean production – provided that clients are ready to pay for the envisaged quality level. Consequently, this business model is dominated by commercial planning, a function that defines products, quality levels and appropriate prices. Network and operations are, rather, cost centres, aiming at delivering the required quality at minimum cost.[3]

Notes

1 Franke, 'Network design strategies'.
2 Franke, 'Network design strategies'.
3 Franke, 'Network design strategies'.

3 Key performance indicators for successful network design

As for every other entrepreneurial or managerial activity, what is key is measuring the success of the decisions and actions taken in the context of airline network management, in order to learn from errors and establish a continuous improvement process. Irrespective of the concrete business model, route networks represent a huge proportion of an airline's revenues and cause a major share of the operational costs. Consequently, *network* performance indicators are *airline* performance indicators.[1]

The definition of success of course depends on the underlying system of objectives, which is first and foremost determined by the ownership structure of the airline. Despite the ongoing trend of deregulation and privatisation, many airlines are still traditional 'flag carriers', wholly or partially owned by the government. In this case, profitability is usually not the top goal of the stakeholders; instead, the creation of jobs, the reliable connection of the country to the rest of the world, or simply prestige happen to be on top of the priority list. Although 'prestige' looks like an unreasonable goal from the perspective of a private entrepreneur, furthering jobs and international air transport can still be very rewarding objectives from a macroeconomic standpoint.

Non-governmental organisations such as the World Economic Forum (WEF) have conducted and published various studies over the past 15 years that underline the positive cycle initiated by an efficient national air transport system.[2] Connecting a country and its citizens (business as well as leisure travellers) to the rest of the world fosters education and wealth; higher education and wealth levels boost, in return, the desire and ability to travel to foreign countries. The wealth-creating effect of an efficient national air transport sector becomes particularly obvious in the case of emerging countries; for highly industrialised countries, the effect is also prevalent, but less decisive in relative terms as a result of the size and maturity of the respective national economies.

Private owners are, of course, rather interested in profitability. Historically, not all of them have acted rationally in this respect, so the prestige of owning an airline seems to be an incentive not only for public but also for private stakeholders. It needs to be assumed, however, that the majority of private owners strive for long-term value creation and profit generation, not for idealistic goals. This is particularly true for airlines that are publicly listed, so belonging to shareholders and institutional investors.

As stated earlier, there is no simple solution to the complex mathematical optimisation problem posed to airline managers by network structures. Consequently, simplified heuristics ('rules of thumb'), as well as key performance indicators (KPIs) play an even bigger role than in industries with linear production systems. Heuristics and KPIs provide a solid foundation – based on experience and best practices – for decisions to be taken in designing and operating a network in the absence of exact mathematical solutions.

3.1 Key performance indicators

The following list shows an overview of the KPIs broadly used in airline network management:

- SLF (seat load factor)
- BESLF (break-even seat load factor)
- Earning capacity
- Yield
- ASK (available seat kilometres) or ASM (available seat miles)
- RPK (revenue passenger kilometres)
- RASK (revenue per ASK) or RASM (revenue per ASM)
- CASK (cost per ASK) or CASM (cost per ASM)
- Aircraft utilisation

3.1.1 Revenue-related indicators

The *seat load factor* (SLF) describes the utilisation of available seats on a certain flight by (paying) passengers. Besides this single flight level, an SLF can also be calculated for more abstract levels, such as (sub-)fleets, sub-networks, all flights to/from a certain market, etc. A good load factor indicates a sustainable utilisation of the resources an airline needs to deploy, especially its seat capacity. Since aircraft are a very expensive asset, and seats in an aircraft represent a perishable capacity, a high SLF is by and large key to achieving overall profitability.

The question, however, is what 'high' means. This question is addressed by the concept of *break-even seat load factor* (BESLF). A seat load factor higher than the BESLF for a certain flight promises – *ceteris paribus* – a positive operating result for that flight. The BESLF is a function of the deployed aircraft (with its specific cost position) as well as of the revenue expected for that flight (depending on fare level and fare mix across available booking classes). Thus, the BESLF is different for each and every flight, though there are typical ranges of BESLF for short-, mid- and long-haul flights (see section 3.2 later in this chapter).

The importance of filling the available capacity of a certain aircraft with paying passengers is also reflected in the term *earning capacity*. An airline can decide to equip a certain aircraft type with more seat rows than other airlines, resulting in more paying passengers per square metre and per aircraft. This approach has been extensively deployed by low-cost carriers, but it has also been adopted by full-service carriers – at least on short-haul flights.

It must be kept in mind, however, that more paying passengers do not automatically translate into more revenue or more profit, since a shorter distance between seat rows ('pitch') reduces the comfort for the passengers, and may thus come at the price of reduced *yield* (fare level/revenue per seat, determined by elasticity of demand). If the fare level is too high for a certain product quality, measured, for example, in seat comfort or pitch, the SLF will probably go down, depending on what competitors do on the same route. The trade-off between earning capacity, yield and SLF is a core equation that every airline and every network manager must deal with to achieve short- and long-term profitability.

The KPI *yield* is always a relative parameter, comparable to 'unit cost'. Besides 'revenue per seat' (see previous paragraph), it can also refer to 'revenue per RPK', 'revenue per ASK' (see below) or 'revenue per flight'. High yields (per seat) are expected to sum up to economically successful flights, and, finally, to a profitable network, provided that the costs are under control.[3] As a consequence of the above-mentioned trade-off, a high yield per seat or per RPK can, but does not necessarily have to, result in the maximum revenue or even profit for a certain flight.

3.1.2 Distance-related indicators

Cost and thus fares on a certain flight are closely correlated with the covered distance (stage length); a long-haul ticket will usually be more expensive than a short-haul flight, since the operating costs – especially fuel – are much higher. This correlation is not linear, though, but degressive, since operating longer routes creates considerable economies of scale. For instance, flying ten times further will not cause ten times higher cost, since by far the largest proportion of fuel consumption occurs during take-off and landing, while flying at cruising altitude and speed consumes much less fuel per kilometre or mile.

In any event, distance is an important factor in analysing KPIs, so there are dedicated parameters taking the mileage into account. The first describes the distance-related *capacity supply* on a flight, route or (sub-)network, called *available seat kilometres/miles* (ASK/ASM). For one flight, this is simply the number of available seats, multiplied by the distance covered during this flight. Aggregated across several flights, this KPI describes the overall distance-related capacity offered by an airline on a certain route, in a certain market or in a (sub-) network. Some European airlines use the term 'SKO' (seat kilometres offered) instead of ASK.

The second distance-related network KPI is the RPK or RPM (*revenue passenger kilometres/miles*). This term must not be confused with the *revenue per ASK* (RASK), explained below. In fact, RPK is not a revenue parameter at all, but multiplies the number of (paying) passengers by the covered distance, determining the distance-related demand for a certain flight. In this context, the term '*revenue* passengers' stands for '*paying* passengers' – as opposed to passengers travelling for free (such as airline employees, or passengers with redemption tickets from loyalty programmes).

For a single flight, the quotient of RPK and ASK is, obviously, equal to the SLF, since, after cancelling the distance from the fraction, the remaining equation is paid seats divided by available seats. For larger entities, the quotient describes a 'system' load factor, which is rather theoretical, but still helpful for analysing the profitability of a route, market or (sub-)network.

The RASK/RASM term is one of the possible yield definitions as explained above. Analogous to the *unit cost* concept, revenue per ASK could be interpreted as *unit revenue*. Since revenue is generated only by paid seats, not by offered seats, an alternative definition such as 'RRPK/RRPM' (revenue per RPK/RPM) would make sense as well; RASK/RASM have achieved much broader recognition as KPIs, however. The difference between RASK and CASK can ultimately be interpreted as the profit per supply volume.

3.1.3 Cost-related indicators and asset utilisation

The *aircraft utilisation* KPI indicates that, even if a flight is fully booked (with an SLF of 100 per cent), this is not yet sufficient to ensure that the capital locked into the aircraft pays off. Unproductive times on the ground reduce the profit of an airline, since passengers pay only for travel time in the air, while many of the costs associated with merely owning an aircraft continue to occur even if it stands still. Thus, an aircraft can definitely cause losses despite sustainable seat load factors if it does not achieve sufficient daily operations times, standing idly on the ground for too long.

The utilisation (i.e., asset productivity) of an aircraft is measured in block hours per day, and is driven by the following parameters:

- the aircraft type (long-haul versus short-haul);
- the business model chosen by the airline;
- the size/scale of the airline; and
- the efficiency of ground handling processes (of airline, airport or third-party service providers).

Aircraft types used for long-haul flights (mostly 'wide-bodies', with two aisles in the cabin) achieve excellent productivity much more easily than those deployed on shorter routes ('narrow-bodies', with only one aisle), since the flown distance is much longer. This is true in absolute terms, but also relative to unproductive times, such as taxiing or waiting for feeder flights with transfer passengers. Whereas wide-body aircraft easily achieve 14 to 16 block hours per day on long-haul routes (with some one and a half flights per day), narrow-body aircraft usually end up with eight to 12 block hours per day (with four to six short-haul flights per day). In this context, low-cost carriers tend to achieve higher aircraft utilisation figures than hub and spoke carriers, since they avoid connecting passengers and thus unproductive waiting times on the ground. Other important drivers of asset productivity are airspace availability and stable operations. Holding patterns, as well as external or internal irregularities, massively compromise aircraft utilisation.

The transfer aspect hints at another major driver: aircraft productivity is dependent on the chosen *business model.* Although traditional hub and spoke carriers operate at large central airports ('hubs') to connect their passengers to more destinations worldwide, LCCs usually avoid the complexity of hubs, and do not support their customers in connecting flights.[4] As a result, LCC flights do not have to wait for feeder flights, which ensures that aircraft can depart again as soon as they have been serviced and filled with new passengers. For LCC flights, the unproductive time on the ground is only determined by the minimum turnaround time (MTT) of the airline at a certain airport, plus the size of the deployed aircraft.

The *size and scale* of an airline exert an impact on aircraft utilisation as well, since it is much easier to generate efficient schedules with big fleets and many rotations, while a schedule with only a few aircraft will usually result in higher production slack and thus poorer productivity in terms of resources.

Finally, the *efficiency of ground handling processes* is also a driver, since slow processes tend to increase the unproductive ground time. Airlines – especially low-cost carriers, which do not enforce waiting times between inbound and outbound flights – do everything to keep the turnaround time of their aircraft as short as possible (at least for short- and mid-haul flights). Consequently, airports advertise their ability to generate quick processes both for aircraft – minimum turnaround time – and for transfer passengers, referred to as minimum connecting time (MCT).

To make managing an airline network even more complex, the nine KPIs mentioned above are partly interlinked. As an example, it is very difficult to achieve both a high seat load factor and a high yield per seat unless the respective market is underserved and short of supply[5] (cf. profitable ranges of SLF, in section 3.2).

3.2 Heuristics and rules of thumb

In non-monopolistic, fully served markets, it is usually very difficult to achieve optimum values for all the KPIs introduced previously. For instance, high seat load factors do not coincide too often with outstanding yields. One airline alone could easily shorten its capacity until the resulting scarcity makes the prices rise, but competitors will gratefully absorb this capacity drain, pushing the equilibrium point back to good load factors and medium yields.

3.2.1 Pricing

A good balance between demand and supply can be achieved through pricing, but also through varying the size (i.e., seat capacity) of deployed aircraft in a certain market. Increased fares usually lead to a reduction of demand, which may result in higher revenues if demand proves to be inelastic or modestly elastic. Vice versa, a highly elastic demand forces airlines rather to lower prices and benefit from increased volume. In both cases, the described effect can be realised only if all competitors play the same game.

3.2.2 Trade-off between aircraft size and frequency of flights

Aircraft capacity is also involved in another key trade-off of network management: the balance between aircraft size and number of services per day. Let us assume for a moment that airline A owns a market share of 10 per cent in a certain market (on a single route), translating into 600 passengers per day. Airline A can deploy different narrow-body aircraft types, featuring 120 and 150 seats. The larger aircraft offers scale effects and thus a better unit cost position (i.e., CASK), so it seems to be smart to deploy *four* services per day ('frequencies') with that larger aircraft, resulting in a 100 per cent SLF, as well as minimum unit cost.

It may still be smarter to offer *five* daily flights with the smaller aircraft type instead, however. The incremental daily service may make the offering significantly more attractive, especially for business travellers, alienating many of them from competitors. Thus, the additional revenue from the improved perception of the service level may more than compensate for the downside of the smaller aircraft in terms of unit cost. This phenomenon is called the 'S-curve effect' (for details, see Chapter 6, section 6.3).

In the absence of reliable calculations, network management practitioners rely on their wealth of experience, providing them with robust assumptions about viable or profitable ranges of KPIs, as well as about the 'sweet spots' of the various trade-offs between value drivers. These heuristics are in many cases absolutely sufficient to solve limited network problems, or to perform a quick reality check of a certain network scenario. The following sections contain a selection of popular and helpful rules of thumb relevant for a variety of network-related questions.

3.2.3 What is a desirable seat load factor?

As mentioned earlier, the profitability of an airline strongly depends on its ability to fill its expensive assets. Historic airline bankruptcy scenarios have very often been scenarios with poor seat load factors, as in the case of former Swissair. In other words: good load factors are not a guarantee for profitable operations – but poor load factors are the safest way to bankruptcy. The break-even seat load factor is a function of many different parameters characteristic of a certain aircraft and route. Nevertheless, there are typical ranges of load factors leading to profitability or unprofitability. On most routes and in most competitive settings, it is very difficult to be profitable with a load factor below 70 per cent – with a few exceptions proving the rule. On short- and mid-haul routes, this threshold is, rather, 75 per cent, since average yields are lower, and unit cost higher, than on long-distance flights. And, finally, low-cost carriers try to achieve a minimum SLF of 80 to 85 per cent, since their unit cost are low but their yields are, proportionately, even lower.

Surprisingly, there is not only a lower but also an upper threshold for profitable load factors. In theory, a seat load factor of 100 per cent would be ideal, and thus highly welcome. In practice, a very high SLF can be counterproductive, at least in non-monopolistic markets with non-scarce supply. In such a setting, it gets more and more difficult to fill an aircraft beyond an SLF of some 90 per cent. In most cases, the last few seats will sell only at discounted rates. Initially, this will simply spoil the incremental revenue of the incremental seats, but, as soon as clients understand the pattern, the total revenue of the flight may get diluted because of dropping prices.

This phenomenon leads to the paradoxical effect that an aircraft with a 98 per cent load factor may contribute *less* total revenue than the same aircraft on the same route on the same day with 'only' a 95 per cent load factor. This is the reason why airlines very often intentionally leave the last few seats empty, as selling them would worsen, not improve, the commercial result of the flight. Consequently, network managers should be cautious as soon as the forecast load factor exceeds 95 per cent. Nota bene, if the market environment is favourable, a 100 per cent load factor can be absolutely perfect – but, in a less supportive setting, it may dilute value.

3.2.4 How many passengers does an airline need on a certain route to justify a direct connection?

Historically, hub and spoke carriers operated their networks with limited risk, since every major airline pursued the same strategy. If possible, they cumulated demand in their hubs, connecting as many origin and destination combinations as possible via that hub, and offering point-to-point services only as an exception (e.g., on trunk routes with huge demand, or on monopolistic routes with superior yield). Although most travellers prefer direct, non-stop connections, they accept a transfer if there is no alternative in the market.

The market entry of low-cost carriers between 1980 and 2000 completely changed the rules of the game in that respect. Since they offered nothing else than point-to-point services, this put massive pressure on hub and spoke carriers to match that offering at least partly. This raised the question of what the minimum market size for a point-to-point service could be. The rule of thumb for this question looks as follows:

Minimum O&D size = average seat capacity x average SLF x average frequency p.w. x 52 weeks

For instance, for an average narrow-body or wide-body aircraft, this equation could have the following results:

Minimum O&D size (narrow-body) = 160 x 0.75 x 5 x 52 = 31,200 passengers p.y.

Minimum O&D size (wide-body) = 400 x 0.70 x 5 x 52 = 72,800 passengers p.y.

The average frequency per week is the result of another heuristic, which is explained below. The estimates of course refer to the passengers a certain airline can attract – not to the total O&D size on a certain route. LCCs tend to need fewer passengers than hub and spoke carriers to sustainably offer direct services, since they benefit from the effect of latent demand. This demand will remain in the head of potential travellers until the price is so low that they are ready to realise their travel wish. For example, after entering the Geneva market in 2003, easyJet boosted overall demand by no less than 70 per cent – so demand for air traffic was obviously 70 per cent higher than observable until that day, just waiting for lower prices to get unleashed. A low-cost carrier could thus start offering point-to-point services even if its tangible O&D share was initially smaller than 30,000 passengers per year – counting on the latent demand effect as a result of offering low fares.

3.2.5 How many passengers are needed to fill one incremental aircraft?

If airlines buy or lease a new aircraft, they usually do so to replace an older aircraft type ('roll-over') or add the capacity to accommodate overall growth plans. Each aircraft needs to be filled to earn money, though, so the incremental perspective is a fair health check of each fleet expansion. For wide-body aircraft, the question is even more critical, since many airlines (especially in the United States) buy or lease new aircraft for dedicated (long-haul) routes.

Minimum passenger number needed to fill an incremental aircraft = average seat capacity x average SLF x number of operational days p.y. x average number of flights p.d.

For average aircraft sizes, here are possible results of the heuristic:

Minimum passenger number (narrow-body) = 160 x 0.75 x 300 x 6 = 216,000 passengers p.y.

Minimum passenger number (wide-body) = 400 x 0.70 x 300 x 1.5 = 126,000 passengers p.y.

It seems to be counterintuitive that there are fewer passengers needed to fill the larger (wide-body) aircraft, but the higher number of daily flights achievable on short-haul routes more than compensates for the smaller seat capacity of a narrow-body compared to a wide-body aircraft.

3.2.6 How many services per day or per week are needed to be profitable on a certain route?

At first glance, there seems to be no direct correlation between the number of flights and the profitability on a certain route, because each and any service should be profitable by itself. Nevertheless, the 'S-curve effect', mentioned above, already hints at the fact that the perceived attractiveness of a schedule is an important efficiency driver. If an airline offers significantly fewer services per day or per week on a certain route than its competitors, travellers – and in particular business travellers – will perceive the service quality of that carrier as inferior to its competitors, and stay away from it.

In smaller markets and/or on long-haul routes, the critical threshold is not the minimum number of frequencies per day (cf. S-curve effect) but the minimum number of services per week. There is statistical evidence that every schedule offering fewer than four services per week will not be perceived as a reasonable offering by travellers, and thus will be ignored. The only exception from this rule are small markets with monopolistic structures. Table 3.1 shows typical ranges of minimum services per week for different market sizes.

It must be kept in mind that four is the absolute minimum for profitable operations, and then only on small O&Ds with limited competition. A service per weekday (i.e., five flights per week) is in most cases the safer bet. As soon as the O&D gets slightly bigger, and attracts more business travellers, a daily service at least (i.e., seven flights per week) becomes more desirable.

All rules of thumb introduced in this section are helpful for average airline and market settings. If the parameters of a certain airline or a certain O&D deviate significantly from the assumptions described above, the respective hints will become invalid. One prominent example is the network effect of feeder flights operated by hub and spoke carriers: in many instances, these airlines will operate flights even though they seem to be unprofitable (e.g., due to poor load factors or low yields).

These flights may bring a considerable number of passengers to the airline's hub, however, where they use long-haul connecting flights with attractive yields. Thus, the feeder flight may look unprofitable by itself, but still create a positive profit effect in the entire network. Opposed to this, low-cost carriers avoid connecting passengers, so an unprofitable flight remains unprofitable irrespective of the rest of the network – and will be eliminated after a short while.

Table 3.1 Minimum profitable number of weekly flights based on experience[a]

	Small O&D	*Mid-sized*	*Large*	*Trunk route*
Short-/mid-haul	4	10...14	15...21	> 28
Long-haul	4	5...7	7...14	15...21

[a] Franke, 'Strategic network management'.

Furthermore, heuristics cannot replace concrete mathematic calculations in designing efficient networks, but they are in any case helpful as quick estimates or plausibility checks for network-related decisions.

3.3 Profit calculation scheme

Proving efficient network design, and taking core decisions in a network, requires of course more than KPIs and rules of thumb. Ultimately, a purposeful and well-structured profit calculation scheme is indispensable for steering the commercial results of a transportation network. The challenges of a profit calculation scheme for airlines are different from those used in industries with linear production systems, though. There are two major concepts of profit calculation that are prevalent in almost every industry.

- The distinction between variable and fixed costs: what are the incremental production costs of one single item, and how much will it contribute to cover the fixed cost?
- The distinction between direct and indirect costs: what is the best way to allocate indirect cost to the production process of a certain product, to measure its overall share in cost and contribution?

In non-linear production systems such as transportation networks, these questions are less clear than in linear systems. In manufacturing cars, for instance, it becomes quite obvious which cost would disappear if the production line stopped, and which cost would remain. In contrast, it is difficult to define for an airline which costs are variable, and which are fixed: variable or fixed relative to what? Another complication is the effect of the remaining time until a certain flight departs. Some design parameters of a network may be absolutely variable years or many months prior to execution (= flight event), but strictly fixed in the short run. For instance, it is easy to lease (or even buy) a new aircraft a couple of months prior to a scheduled flight, but this will usually not be taken into consideration if only a few weeks or even days are left.

Thus, most airlines deploy a staged *profit contribution calculation scheme* with the following steps.

- *Long-term perspective* (> 1.5 years prior to flight event): strategic time horizon; all assets and resources are basically variable, especially the fleet size and structure.
- *Mid-term perspective* (> 0.5 years, < 1.5 years prior to flight event): scheduling time horizon; fleet is mostly fixed, but selection of routes and allocation of resources to flights is still variable.
- *Short-term perspective* (< 0.5 years prior to flight event): schedule adjustment time horizon; fleet and schedule are mostly fixed, but allocation of clients to flights is still variable.

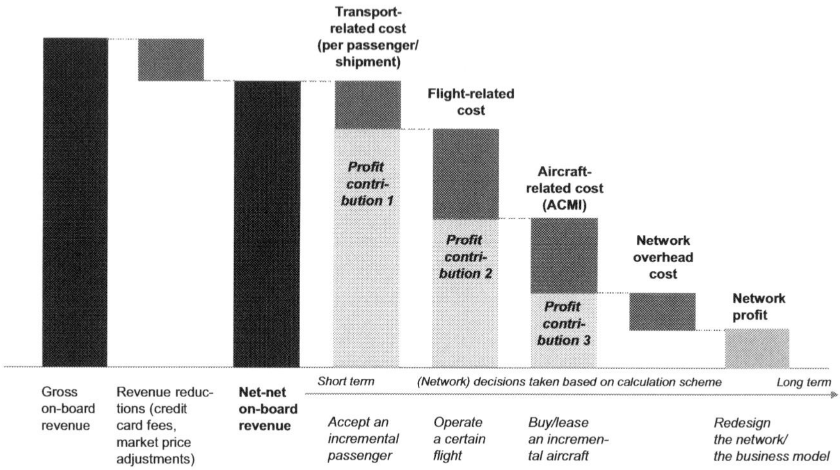

Figure 3.1 Network profit calculation scheme[a]

[a] Franke, 'Strategic network management'.

The profit calculation scheme proceeds in reverse order – i.e. from short-term decisions to long-term decisions – as depicted in the waterfall chart in Figure 3.1.

The starting point of the cascade of costs and profit contribution stages is the gross revenue determined by ticket sales. This figure is then adjusted by accounting for sales reductions (e.g., credit card provisions) and market price adjustments (e.g., kickbacks or net fares for corporate clients). The result is called net-net revenue (two times 'net' for two adjustments), representing the effective revenue actually available for covering variable and fixed costs.

3.3.1 First stage of the network profit calculation scheme

In the *first stage* of the calculation scheme, only those costs that can still be influenced in the short run are taken into account. Under the assumption that fleet and schedule are mostly fixed, the remaining decision that can be taken at short notice is whether or not an incremental passenger should be accepted for that flight (provided that seats are still available). Expenses related to this instance of the scheme are called '*transport-related costs*' or '*passenger-related costs*'. Variables on this level are direct on-board costs such as catering, passenger-related insurances and the handling and security fees per passenger occurring at the origin and destination airports. Furthermore, each passenger causes incremental fuel consumption, as well as incremental take-off and landing fees at the airports (since partly based on the weight of the aircraft); most profit schemes simplify the calculation, however, by pushing the latter two aspects to the next stage.

The difference between the net-net revenue and the passenger-related costs is called *profit contribution 1* (or contribution margin 1, in relative terms – i.e., as a percentage figure in terms of the respective revenue), indicating which part of the revenue is available to cover schedule and fleet costs. If the fare expected from the incremental passenger is smaller than the passenger-related cost, this passenger should not receive a seat on the flight.

3.3.2 Second stage of the network profit calculation scheme

The *second stage* of the scheme deals with the mid-term time horizon, when the fleet is mostly fixed, but the schedule (i.e., flights, rotations) is still flexible. The dominant decision to be taken at this stage is whether an envisaged flight on a certain route should be performed as planned or, rather, cancelled. The expenses on this level are called '*flight-related costs*', and they comprise the direct costs of a flight, such as fuel consumption, air navigation fees, ground handling fees and aircraft-related airport fees (for take-off, landing and parking). Furthermore, complex flight patterns/rotations may cause above-average crew costs (reflecting overtime payments, dead-end flights, overnight stays), but, as with the previous stage, these costs are usually pushed to the next stage for the sake of simplification.

Deducting flight-related costs from profit contribution 1 leads to *profit contribution 2*, available to cover all subsequent direct and indirect costs. If the overall revenue from a certain flight is smaller than the flight-related costs, this flight should be cancelled – at least, if this turns out to be structural loss. For hub and spoke carriers, however, even a persistently negative profit contribution 2 may be acceptable, provided that this flight contributes to disproportionately larger profits on subsequent (long-haul) flights. In this case, the positive network effect through high-value transfer passengers beats the negative profit contribution of the flight itself.

3.3.3 Third stage of the network profit calculation scheme

The *third stage* of the profit calculation scheme covers those costs that can be influenced only with a considerable lead time, related to a strategic perspective. The main resource in the focus of this stage is the fleet. Consequently, the variable costs at this stage are the so-called '*aircraft-related costs*'. Another very popular term for this cost block is *ACMI*, an acronym for 'aircraft, crew, maintenance, insurance', hinting at the major cost blocks on this level. *Aircraft* relates to the financing or leasing costs, *crew* to the salaries of pilots and flight attendants, *maintenance* speaks for itself, and *insurance* refers to aircraft-related insurance rates.

ACMI is also the name of a frequent airline business model: smaller airlines build up efficient operations, and, instead of marketing their seats themselves, they lease their capacity to larger airlines. This may add value if the so-called ACMI carrier is so efficient (e.g., on account of low crew salaries and lean operations) that its specialisation advantage is larger than the lost scale effect of the receiving airline – at least, for a certain sub-network. The receiving airline has the choice of leasing the aircraft of the operator including the crews ('wet lease') or without them ('dry lease').

The difference between profit contribution 2 and aircraft-related (ACMI) costs is called *profit contribution 3*, and determines the profit share that remains to cover all the indirect and overhead costs of operating a network – such as the salaries of professionals planning the network, or managing ticket sales. If the incremental revenue envisaged for a new aircraft is smaller than its ACMI costs, this aircraft should not be purchased.

The residual value after deducting indirect and overhead costs from profit contribution 3 is called *network profit*, and represents a major share of the overall operational profit (earnings before interest and tax: EBIT) of an airline. If the network does not create a profit, but a loss, the overall network structure and business model need to be carefully examined – and, potentially, adjusted.

3.3.4 Impact of fuel price and other external factors on airline profit

The relative share of passenger-, flight- and aircraft-related cost varies widely, depending on drivers such as:

- the business model operated by the respective carrier;
- scale effects (e.g., size of fleet, etc.);
- the actual fuel cost level;
- the salary level and labour laws in the home country of the respective carrier; and
- the fee level at home airport(s)/hubs.

For instance, the global oil price (driving the price of aviation fuel) has oscillated for the past 20 years, between US$19 per barrel (WTI/Brent Crude) in January 2002, US$140 per barrel in June 2008 and US$22 per barrel in March 2020,[6] featuring growth by a factor of 7.4 within only six years, then a decline by a factor of 6.4 over the next 12 years.

Whereas fuel cost (as a major part of the flight-related costs) accounted for some 10 to 15 per cent of the entire network costs of a hub and spoke carrier back in 2000, this proportion grew substantially till 2011. Under the assumption that all other cost elements stagnated, growth by a factor of 5.4 would have led to a fuel cost share of 40 per cent by 2011. The only two reasons why this figure did not become effective at that time were, first, the increase of other cost elements and, second, the fact that most major airlines secured lower fuel prices for quite some time through forward contracts ('hedging'), which saved them substantial amounts of money around the end of the first decade of the twenty-first century. Hedging helped them limit the fuel cost share to some 22 to 28 per cent at the beginning of the financial 'subprime' crisis in 2008/9 – a range that is still quite common ten years later.

After the financial crisis the boom in fracking technology pushed the oil price down to a range between US$50 and 70 per barrel, plummeting further, towards US$22, in spring 2020 because of the global COVID-19 crisis. In parallel, the fuel cost share continued to diminish towards the level that it used to have 20 years ago (unfortunately, this is a theoretical value, since airlines worldwide had to stop most of their operations in March 2020 due to the pandemic).

For low-cost carriers, the fuel cost share in general is significantly higher than for hub and spoke carriers, since the former are able to push down all other cost blocks below the level of the latter. Consequently, the fuel share reached a peak of close to 40 per cent of the overall network cost of LCCs at the beginning of the financial crisis.

Keeping the above reservations in mind, there are nonetheless typical ranges of the different cost blocks for hub and spoke carriers:

- transport-/passenger-related costs: 10 to 15 per cent of net-net revenues;
- flight-related costs: 30 to 40 per cent (strongly dependent on the fuel price level, as explained above);
- aircraft-related costs: 30 to 35 per cent (dependent on aircraft ownership model, scale, deployed aircraft types and age of fleet); and
- indirect and network overhead costs: 10 to 15 per cent.

3.3.5 Impact of overcapacities on airline network profit

These figures lead to a typical network profit level of 5 to 10 per cent, which is a mediocre result for an asset-intensive industry such as aviation. In fact, only a very few hub and spoke carriers have ever earned their capital cost throughout the past 100 years, let alone achieve sustainable profitability. This is the reason why aviation is called a '*stalemate*' industry by analysts, which means that all carriers together create no economic value in the long run, and just a very few outstanding airlines earn acceptable profits, at the expense of all other airlines, in the short run. There are three major reasons for this persistent, long-standing phenomenon, which can hardly even be found in any other industry:

- non-profit-oriented shareholders, such as governments (interested instead in prestige, jobs and the wealth-creating effect of air transport);
- rigid regulation (though it has continuously diminished since the late 1970s); and
- low entry hurdles, and high exit hurdles, for airline founders.

Ultimately, these three reasons merge into one mega-reason: the perpetuation of massive overcapacities in the air transport market. Once an aircraft has been purchased and commissioned for service, its capacity will not disappear from civil aviation for the next 20 to 30 years. Regular maintenance will keep it perfectly safe throughout this period, and obsolescence effects will be compensated for by low ownership costs after the aircraft is fully depreciated.

3.3.6 Profitability of low-cost carriers

Low-cost carriers have been successful in achieving good profits against the industry trend, however, as explained above. Leading carriers in this segment, such as Ryanair, Southwest Airlines and easyJet, have historically posted profit margins of 15 to 20 per cent, which means that they cover their capital cost, and create economic value. The obvious reason for this success lies in the superior unit cost position of these airlines, generated with the help of a truly lean production and overhead model, as well as a very low level of complexity. There have, of course, been dozens of unsuccessful and unprofitable low-cost carriers around the globe, but these carriers have gone bankrupt and vanished from the industry (as opposed to their aircraft). This hints at another major difference between low-cost carriers and hub and spoke carriers: not only are they cheaper, providing them with a larger leeway to generate profit, but they are also owned by more rational investors; if a route does not generate profits, they cancel it, and, if the airline as a whole does not earn its capital cost, they liquidate it.

There have also been a few hub and spoke carriers posting operating profit margins above 10 per cent, but this has been in 'golden times', such as the second half of the 1990s or 2006/7. Furthermore, above-average profit margins can be achieved in the short and medium term by postponing required investments, especially into the recurring renewal of the fleet. Pushing this renewal to the future means boosting profits today at the expense of poor profits tomorrow; this approach will backfire at a later stage.

There are other important drivers of network profit, such as aircraft type and size, frequency of services and hub patterns (for network carriers). These drivers are, at the same time, network design and planning parameters, however. Thus, they will be discussed in the next chapter. A good overview on airline economics and management can be found by reading Doganis.[7]

Notes

1 Franke, 'Network design strategies'.
2 Blanke, J., and Chiesa, T., *The Travel and Tourism Competitiveness Report 2007: Furthering the Process of Economic Development*, Geneva: World Economic Forum, 2007.
3 Franke, 'Network design strategies'.
4 Franke, 'Network design strategies'.
5 Franke, 'Network design strategies'.
6 See https://tradingeconomics.com/commodity/brent-crude-oil.
7 Doganis, R., *Flying Off Course: Airline Economics and Marketing* (4th edn), London: Routledge, 2009.

4 Network management processes

4.1 Interplay between network management processes, accountabilities and tools

As with every other managerial function, exerting the function in daily business requires an interplay between business processes, organisational accountabilities and (IT) tools. Or, in other words, a suitable operational and organisational structure, supported by appropriate systems. For complex tasks such as (airline) network management, there is usually a certain offering of IT tools in the market, such as scheduling tools, rotational planning systems, hub optimisers and fleet allocation tools (cf. Chapter 5). It is seductive to buy one of these tools (or even integrated IT suites) and build respective accountabilities and processes around them (‘*process follows ownership follows tool*’).

This approach may work for small tasks, but it is highly critical for major functions such as network management. The reason for this lies not just in the complexity but also in the required variability of network management. If two similar airlines design and operate their networks, the results may be similar as well – but they can also be quite different.

The reason for this counterintuitive phenomenon lies in the importance of the internal and external framework of the two airlines, impacting the efficiency of a certain network solution. For instance, operating hours or night curfews may be different between the hubs of the two airlines, as may be the labour laws of their home countries (external drivers). Likewise, the ideal traffic structure in the two hubs (‘banks’) will not be equal, since the directionality and rotation duration of the covered destinations happen to differ significantly. Or, on the asset side, fleet composition and age – exerting an impact on unit cost and route allocation – will differ as well (internal drivers).

Consequently, even with similar business models and the same network tools, the outcome of the network design or planning exercise should be different. In an ideal world, design and planning processes appropriate for a certain (network) environment should entail both accountabilities and tool requirements (‘*tool follows ownership follows process*’). The downside of this puristic approach is the underleverage of tools well established in the airline IT market; a considerable number of functions of a popular and proven IT suite may not be usable if processes are customised perfectly for a certain airline strategy or setting. Or, even worse: the most common IT tools in the market cannot be deployed *at all* at that airline, since they do not really support the chosen process structure; in this case, a customised set of tools needs to be developed, with all the costs and risks inherent in such an endeavour.

The upside, on the other hand, is that the airline does not have to adapt its delicate processes to the generic tool – with all the potential losses and cut-offs – but can adapt the tool support to the customised processes and accountabilities. As always, however, there is rarely just black and white, but a lot of grey in between; decision makers thinking about this trade-off should have a very close look at network design and planning tools available in the market, and ponder the benefits of a customised or even self-developed IT suite against the risks and extra costs of this puristic approach. A compromise or hybrid solution is in any event acceptable if the ideal solution entails too much risk.

4.2 Integrated versus fragmented network management approach

As stated earlier, there is no fully integrated optimisation model or network solution so far – partly because of limited computational power, partly because of the different conditions under which airlines operate their networks (see above). Furthermore, this model would have to be suited to plan and steer the involved resources in all instances, from strategic scenarios all the way to the operational allocation of resources, which is a quite unrealistic specification.

In the absence of a fully integrated tool, airlines deploy a series of sequential optimisation steps, unravelling the mathematical problem to a level at which a subset of the problem can be handled by state-of-the-art algorithms and computers. There are two dimensions of unravelling: by *type of resource* and by *timeline*[1] (see following sections 4.3 and 4.4).

4.3 Unravelling network management processes by type of resource

Network management deals with allocating resources to transportation processes in the most efficient way. In a fully integrated approach, a single optimisation model would do that simultaneously for involved assets and resources. As stated earlier, this integrated model does not exist today, and it would not really be wise to use it, since every airline has its specific set of strategic goals, environmental factors and local operational limitations. Furthermore, from the perspective of the assets to be allocated, it is not really *necessary* to plan all resources at the same time.

The reason for this is the fact that the resources deployed by an airline are enormously heterogeneous with respect to their value (i.e., capital lock-up), availability and flexibility. An aircraft is much more expensive than a single crew member or an intermediate maintenance event, and it is much more difficult to exchange it in the short term. Planning, for instance, an aircraft rotation and the respective crew rotations in an integrated manner would suggest that they can also be adjusted in one step – which is not the case. On the contrary, the rotations of crews are subject to different limitations from those of an aircraft, and the change process is different. Thus, what is the value of planning them at the same time in one process?

Consequently, it is common practice not only for airlines but for all transportation companies to pursue the following planning paradigm: first determine those resources that have the highest asset value, or are least flexibly available, then do the same with all other assets or resources in declining order of asset value or cost. For airlines, for example, selecting a hub locks up considerable capital (such as infrastructure) and cannot be revised quickly. Then, aircraft type and deployment patterns will be selected, since aircraft absorb enormous investments (if they are not purchased, then they are leased, with considerable lease rates).

All other resources or assets are determined in the order of either declining capital lock or of increased flexibility, such as crews, maintenance capacities and ground handling staff (see illustration in Figure 4.1). This sequenced approach is a simple heuristic, but it is pragmatic and delivers reasonably efficient network designs.

On a more detailed level, many airlines also split their network management by the type of aircraft required for certain network structures. They design and schedule long-haul networks (served with wide-body aircraft) separately from short- and mid-haul networks (mainly served with narrow-body aircraft), because the mechanics of these layers are quite different, and the interdependences are limited. This approach reduces the immanent complexity of network management for hub and spoke carriers, since the only remaining touch points between the two spheres are the connections in the hub.

4.4 Unravelling network management processes along the timeline

In a complex environment such as the aviation industry, environmental as well as internal factors tend to change before the day of operations and evolve differently from what was planned earlier. Airlines consequently apply a planning sequence with at least three steps:

- long-term (strategic) planning (one and a half to three years up to ten years prior to a certain flight event);
- mid-term market planning and scheduling (three to six months up to one and a half years prior to a flight event); and
- short-term operational planning and resource steering (the last few weeks prior to a flight event).

Figure 4.2 shows an idealised process flow containing these three steps.

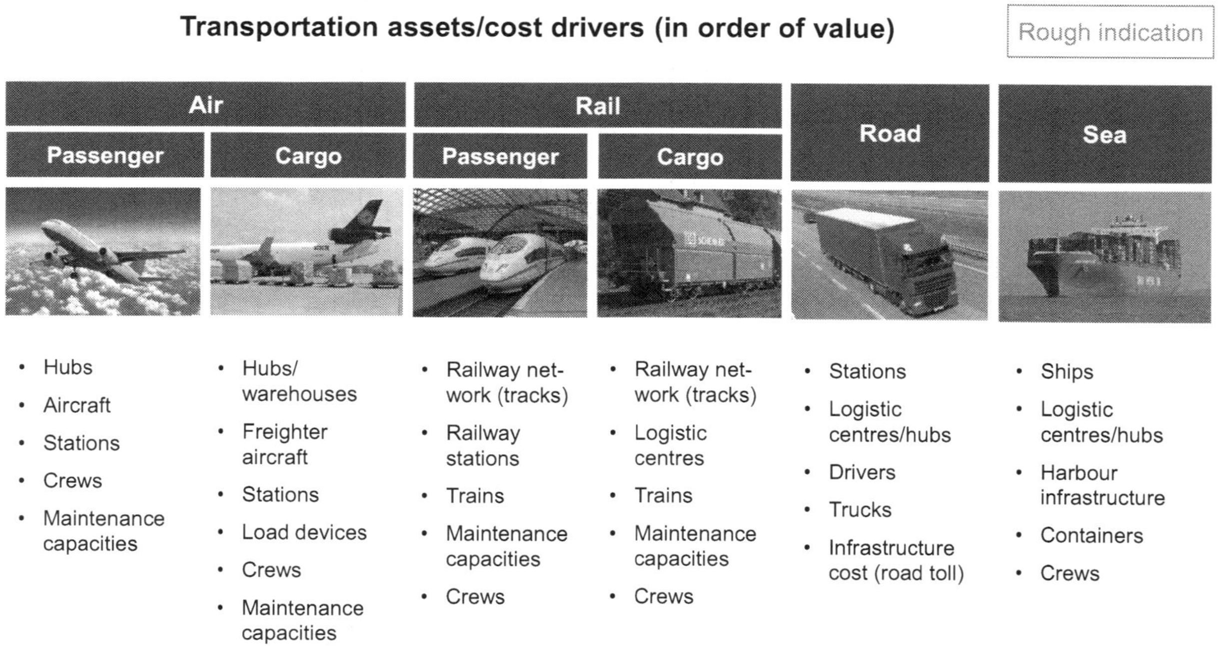

Figure 4.1 List of transportation assets by industry listed in the order of asset cost[a]

[a] Franke, 'Network design strategies'.

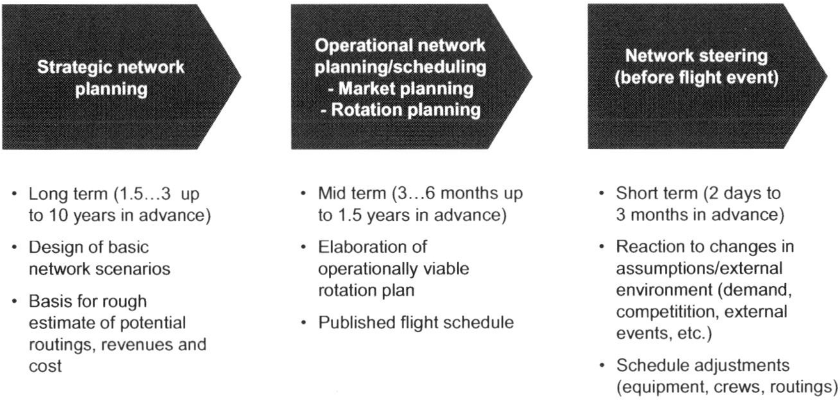

Figure 4.2 Idealised process flow of network planning and steering[a]

[a] Franke, 'Network design strategies'.

4.4.1 Long-term network planning

Long-term planning primarily deals with network resources representing a high value and/or featuring limited flexibility in acquisition or deployment. For instance, an aircraft locks up a sizeable value (depending, of course, on its size and technical equipment), and it may be difficult to buy or lease a particularly popular aircraft type during an economic *upturn*. When every airline wishes to expand its capacities, efficient aircraft types may become scarce, with original equipment manufacturers (OEMs) posting waiting lists of two years or more. Likewise, it is challenging to sell or terminate a leasing contract for an aircraft during an economic *downturn*, at least at short notice. Consequently, it seems to be appropriate to pursue an anti-cyclical fleet renewal policy, though most airlines find it impracticable to sell aircraft in high-demand periods and buy them in times with poor demand.

On the other hand, once an aircraft has been added to an airline's fleet, it is rather flexible to be deployed wherever and whenever it is needed – provided that operational prerequisites such as licensed crew, maintenance capacity and airport and airspace slots are available. Even the seat capacity and seat configuration, defined by cabin classes, can be adapted within certain limits to comply with changed demand, albeit not at short notice.

The least flexible asset of an airline is its hub airport(s). Once a hub and spoke carrier has decided to establish itself at a certain home base, it needs to invest money in making this home base work in a way that is fully in line with the operational and traffic-related requirements of its network. The airline operating this airport as its hub needs to rent or build sizeable premises, and assign significant staff to this location (sales, customer service, passenger handling, resource disposition, etc.).

Furthermore, crews will have this airport as their home base, which fact exerts a considerable impact on the efficiency of the crew planning process. Once an airline has gone down that road, the exit hurdles are prohibitively high, so the airline will try to protect its investments and stick to its hub choice for many years, if not decades. For low-cost carriers, on the other hand, the choice of their home bases is much less binding and long-term than the decision of a hub, so the long-term planning function of a low-cost carrier is rather lean compared to a hub and spoke carrier.

In line with this notion of long-term cost and lacking flexibility, the main task of long-term network planning is the design of network scenarios that leverage the determined resources (home base/hub, fleet) in the most efficient way. This approach is also the basis for the most fundamental network planning paradigm, introduced in section 4.3.

The granularity and composition of strategic network scenarios varies across business models and geographies. For instance, European hub and spoke carriers traditionally put more effort into strategic planning than US carriers. There is a common core in all long-term planning paradigms, though, which is the design and subsequent evaluation of so-called O&D demand bundles. These scenarios are defined on an aggregated level, matching high-level demand forecasts with O&D bundles (e.g., 'Western Europe to East Coast United States').

On this level, network scenarios are not detailed enough to probe operational viability, but suited to roughly calculate passenger and revenue flows, as well as market coverage and fleet utilisation. Overarching financial targets of the airline, such as revenue growth, asset productivity and profitability, can be simulated and adjusted if required. The results of the network scenario calculations will also be used as input for the subsequent steps, such as fleet planning and market planning.

4.4.2 Mid-term network planning

Mid-term planning is the second step of the tripartite network management process flow. Between 18 and six months prior to a physical flight event, airlines translate aggregated network scenarios into concrete schedules ready for client bookings. There are two sub-steps: first, a market plan is designed, featuring services that are expected to be highly compelling for relevant passenger flows. In this context, the term 'service' comprises flights on a certain O&D, at a certain time within a certain frequency pattern and associated with a certain service level. Sophisticated market planning models take the effect of competition into account, comparing the perceived attractiveness of a market plan with that of other airlines competing on respective routes.

These tools, often referred to as market models, define the relative attractiveness of a schedule in a competitive environment, and allocate market shares (with respect to passenger flows) to all carriers operating on a certain route. For instance, a schedule may look attractive enough upfront to gain a market share of some 30 per cent on a specific route, while the market model downgrades this share to 20 per cent after several iterations, deeming the schedule of a competitor more compelling than its own.

4.4.3 Market planning and rotational planning

The market plan consists of concrete O&Ds and real aircraft rotations (as opposed to the O&D bundles used in the long-term planning process), but it does not ensure the operational viability of the schedule. This task is covered by the second sub-step of mid-term planning, which is called 'rotation planning'. *Market planning* identifies potential demand on O&Ds that can be covered by the respective airline and assigns available seat capacity to it. *Rotation planning* translates this rough capacity offering into physically viable roundtrips of aircraft and crews, observing all relevant operational requirements and restrictions. The mission of rotation planning is to make sure that the schedule is entirely feasible across the following dimensions:

- the availability of the required aircraft type on the day of operations (featuring the right seat capacity and flight range);
- the availability of crew (cockpit and cabin), licensed for this aircraft type, on the day of operation;
- the availability of slots (departure airport, airspace, arrival airport);
- the availability of required (line) maintenance capacities at the arrival airport;
- the existence of required regulatory permissions and certificates (e.g., traffic rights, air operator certificate (AOC)); and
- compliance with local/international rules and restrictions (e.g., embargos, night curfews).

Furthermore, rotation planners check whether the asset productivity (i.e., aircraft utilisation) is maximum for the given setting, to ensure commercial viability as well. This aspect is further detailed in Chapter 6, sections 6.3 and 6.5.

Market planning and rotation planning ultimately deliver the final schedule that can be offered to clients – historically as paperback booklets, nowadays only in electronic format (on airline websites, in airline apps or in meta-search engines: third-party websites screening all available offerings for a certain route or market). Internally, the flight schedule files represent the basis for all subsequent resource planning and steering processes, since they embody not only the marketable offering of the airline but also its production programme. A published flight schedule may have different formats, but it commonly contains the following pieces of information:

- origin and destination airport of the flight
- timing (scheduled departure and arrival time, duration)
- flight number (including potential multiple numbers of code share partners)
- aircraft type

- for non-direct flights (transfer traffic connected via the hub of a hub and spoke carrier): location of the hub airport, waiting time in the hub; timing, flight number and aircraft type of the second flight

4.4.4 Customer preferences and buying behaviour

Many clients choose their preferred flight based on the direct versus non-direct information from the list above, since they like non-stop flights better than connected services, perceiving a transfer as an inconvenience. On the other hand, clients accept one-stop flights on long-haul routes since they understand that airlines are economically unable to offer smaller destinations directly (i.e., point to point). Even if there are one-stop flights only on certain O&Ds, airline clients prefer those with the shortest overall duration ('elapsed time') – especially business travellers, who are usually time-sensitive.

Thus, the ability of an airline to ensure short waiting times in their hub(s) for connected traffic is a key differentiator against competition in the long-haul business. This ability is correlated with the peaky traffic pattern ('bank structure') that airlines establish in their hubs to maximise connectivity for their passengers and productivity for their assets. This concept is further detailed in Chapter 6, section 6.5.

There are of course travellers with different preferences and buying behaviours – especially leisure travellers, who are more price-sensitive and less time- or service-sensitive than business travellers (or wealthy private travellers). Consequently, they tend to choose their flight(s) primarily on price and accept longer trip durations or more stops in return. Flight itineraries with more than two stops are highly unpopular for almost every type of traveller, however – unless they are absolutely unavoidable in order to connect a remote place with another remote place on the other side of the globe.

4.4.5 Impact of the geopolitical location of a hub on network efficiency

Some airlines are forced to live with a geographic or operational setting in their hub(s) that makes it almost impossible to ensure short connecting times for transfer passengers. For instance, hub and spoke carriers in Europe, Southeast Asia/Greater China or the US East Coast are located in geographically very favourable positions in the middle of the major traffic flows – occurring mainly in the Northern Hemisphere. Furthermore, these locations usually benefit from moderate weather conditions. Thus, the only obstacles that can prevent the respective carriers from highly productive 24/7 operations, spreading flights across the day and enabling short connecting times for transfer passengers, are regulatory restrictions such as night curfews or air pollution limits.

In other regions, airline rotations do not fit that nicely into the major traffic flows. This is true for extremely remote spots such as Australia and New Zealand, but also for carriers in the Middle East. Whereas European carriers massively benefit from a nice balance between westbound and eastbound flights – in terms of distance, directionality and roundtrip time – carriers in the Gulf region need to accommodate longer westbound flights but shorter eastbound flights in their rotational patterns (outbound; vice versa for inbound flights).

Furthermore, carriers such as Emirates, Etihad and Qatar Airways have built their traffic systems primarily on long-haul flights, since the local catchment is very small; this leads to many fewer connections between short-haul (feeder/de-feeder) flights and long-haul services than is performed by European carriers in their hubs. Finally, Middle East carriers have to live with more extreme climatic conditions, such as very high daytime temperatures and sandstorms, limiting the operating hours of the respective hub.

For instance, Emirates limits the majority of its inbound and outbound flights to a relatively short time window between late afternoon and early morning. Emirates passengers from US and European origins arrive between evening and midnight, then have to wait for four to six hours till their onward flights to East Asia depart in the morning. Bank structures more convenient for transfer passengers would lead to less productive rotational patterns, or to departures during the midday heat – reducing the performance of jet engines and turning aircraft cabins quickly into a sauna.

The respective airports, such as Dubai International and Doha, have tried to react to these downsides, making a virtue out of necessity by building very attractive terminals with a wide range of shopping and relaxation opportunities. The idle waiting time is turned into a pleasant experience – a strategy that has persuaded many travellers to accept longer connecting times.

4.4.6 Role of global distribution systems (GDSs) in airline distribution

The schedule data mentioned above are not only published by airlines themselves but also provide the basis for two very powerful third-party distribution channels: since the early 1990s global distribution systems such as Sabre, Amadeus and Worldspan have achieved a crucial role in airline distribution. Whereas travellers had to extract their preferred travel solution themselves from a variety of separate airline schedules before 1990, the GDS providers independently built huge computer-based inventories, digesting all airline schedules they could get hold of.

Moreover, they built filtering and prioritisation options, mimicking the assumed selection criteria of travellers (non-stop before one-stop, shortest elapsed time, lowest price, etc.). The resulting reservation system turned out to be a cash cow, since it made looking for flights so much more convenient for travellers; GDS providers were able to raise service fees not only from travel agents using the tools, but also from airlines.

Once the GDS channel dominated the distribution chain, airlines could hardly live without it; its global reach and efficiency made it indispensable for almost every player in the market. This did not change until the rise of the low-cost carrier business model in the late 1990s, when these carriers build their distribution channels from scratch and avoided the efficient, but costly, GDS support. Instead, they fully relied on their proprietary online channel, keeping their sales cost very low. Hub and spoke carriers continued to use GDS as major sales channel, but tried in parallel to negotiate more flexible contracts, and support alternative distribution channels (the so-called 'New Distribution Capability': NDC).

The selection mechanism introduced by the GDS providers has been so successful ever since that it has even changed the way airlines design their schedules. Subsequently, a self-fulfilling prophecy was established: GDS providers assume that travellers consider a few dominating decision criteria for booking flights, and turn them into filter criteria. Travel agents and travellers find these criteria intuitive, and select flights accordingly. And, finally, airlines optimise their schedules to best comply with the selection criteria once they become effective in the sales process. For example, airlines may try to depart ten minutes earlier than their competitors on a certain route, or achieve an elapsed time ten minutes shorter, since that will be rewarded by the GDS algorithm with a higher ranking among the listed options. This kind of schedule fine-tuning may look highly overengineered – if not useless – but, as long as the clients reward the current filtering logic, no airline will skip this lever.

The ultimate goal for each airline in this context is achieving a so-called 'first page appearance', i.e. being shown as high as possible on this first page. Statistics prove that the self-fulfilling prophecy mentioned above is still fully functional: the probability of selling a certain flight rises enormously if that flight is ranked among the first three to five postings on the first page of a GDS display. Putting it the other way round, the probability of selling that flight is close to zero if the flight does not make it to the first page.

Traditional hub and spoke carriers and low-cost carriers have adapted their business models to external and internal needs over the past 20 years, coming closer to each other in many aspects, such as lean operations and ancillary services. This is also true for distribution channels, which fact has, ironically, led to a situation in which hub and spoke carriers try to avoid the GDS fees, while low-cost carriers are forced by sheer size and growth to start using GDS services.

4.4.7 Additional operational data in the flight schedule

Besides the schedule data mentioned above, which are relevant for travellers to select their preferred flight, a schedule also contains operational data not published:

- slot information
- dispatch data such as distances, maximum take-off weight (MTOW), fuel consumption
- number of crew members with a certain licence (cockpit/cabin)

- maintenance needs
- expected load factor per cabin class
- regulatory information such as night curfews, overflight bans, etc.

These data are the starting point for any operational steering and disposition process hereafter. For example, a schedule contains assumptions about a suitable aircraft type (in terms of seat capacity, range, cabin configuration and unit cost) as well as about potential crew members with the required licences and training for that type of aircraft. At a later stage the specific aircraft (determined by its 'tail number') and appropriate staff members are selected and assigned to the operational schedule. In this regard, the initial schedule is operationally viable, but not yet sufficient to be physically operated. This last step is done shortly before the flight event.

4.4.8 Frequency of flight schedules

Historically, airlines tended to produce two to four different schedules per year, depending on the seasonality of their demand. In Europe, for instance, it was common to publish a summer schedule, a holiday schedule and a winter schedule. This distinction was closely linked to the slot allocation mechanism in global aviation; since slots are very scarce at some congested airports, the International Air Transport Association (IATA) has established a slot conference twice a year (one around March for the next winter period, the other around November for the next summer period), at which airline representatives post their slot needs for the upcoming schedule period, and a collective allocation process tries to define a fair assignment of the available slots to these requests.

The slot conferences are still active, but the environment has changed greatly, so that most airlines nowadays go for ten schedules or more per year. Higher computational power and process digitalisation have led to a drop in time consumption for a single schedule. The more granular offering creates a better fit with the current travel preferences of clients and is deemed to be worth the effort. The increased frequency of scheduling has also changed the time horizon of each planning step: whereas, in former times, the scheduling started some 18 months prior to the respective flight event, creating a ready schedule at the latest six months before flight, the process has obviously become more frequent and more short-term. With ten or more different schedules per year, the handover into operational steering takes place at a later point in time, around two to three months before the flight event.

4.4.9 Short-term network planning and resource steering

Short-term planning denominates the last and most current part of the overall process flow. In a perfect world the network planning process would be accomplished after scheduling, but in real life there are changes and disruptions preventing the schedule from being executed in the way it was initially planned.

There are external disruption drivers, such as bad weather conditions, political crises or congested airspace, as well as internal drivers, such as technical issues, crew absence or missed connections. There are three triggers for short-term planning or steering.

- External or internal factors change after the schedule has been finalised, but before the flight event (e.g., a short-term drop in demand in a certain market because of a political crisis, or a technical issue of an aircraft type); in this case, schedule elements such as aircraft allocation, the timing of flights and even certain destinations can be adjusted to accommodate the changed environment.
- External and internal factors remain unchanged, but network planners continue to calculate scenarios, and switch resources within the last few weeks prior to the flight event. These deliberate, unforced changes follow the theory that short-term schedule adjustments can still generate incremental value – even though these changes cause a certain instability in the process. A prominent example of this practice was the so-called 'dynamic fleet reassignment' tested by various US carriers, based on a theory developed at the Department for Aeronautics and Astronautics of the Massachusetts Institute of Technology (MIT), based in Cambridge, Massachusetts.
- External or internal factors change within the last 72 hours before departure, or after departure of the flight. A forced change within this timeframe is referred to as a disruption, and needs to be addressed under time pressure. The disruption can be caused by bad weather conditions, congested airspace, sickness of crew members or passengers, delayed ground handling processes, etc. (for further reading about operational disruption management, see Bruce).[2]

The first two cases are handled by rotational planners, whereas the disruption case is a task for the operations control function of an airline. In a stricter sense, the third case is not a planning process any more but the purely reactive generation of solutions for disruptions. The handover between network planning and operations control takes place between 72 and 48 hours prior to departure. In other transportation industries, such as rail and forwarding (the land-based transport of goods), the process of handling disruptions occurring during operations is also called *disposition* – a term that is rather uncommon in aviation.

4.4.10 Cascading network planning processes over time

The three planning processes described above represent an iterative, continuously revolving and cascading flow. A flight schedule is, quite naturally, the result of a preceding design of a strategic network scenario, and, likewise, every short-term schedule adjustment of course builds on the current flight schedule. Since the time horizons and cycle durations of these planning steps differ, however – from several years for a strategic network scenario to several months for a flight schedule, down to a few weeks for short-term adjustments – there is not just *one* linear flow through these three steps, but multiple flows.

Let us assume, for example, that an airline kicks off a strategy process in spring 2020, producing network scenarios for the next ten years (2021 to 2030). Nine out of these ten years have already been reviewed in the course of the last strategic planning process in 2019 (eight in 2018, and so on), so only the assumptions and results for 2030 are brand new. On the other hand, the results for the first nine years can be challenged (based on new insights) and adjusted if appropriate.

In summer 2020 a mid-term planning cycle for the next three years will be launched (2021 to 2023), which can be built on the previous two mid-term planning results (from 2018, for 2019 to 2021, and from 2019, for 2020 to 2022). In addition, this planning effort can also consider the results of at least the latest two strategy planning cycles (from 2018, for 2019 to 2028, and from 2019, for 2020 to 2029). Each mid-term planning exercise can be checked for consistency *vertically* (against all comparable preceding strategic planning results) and *horizontally* (against all comparable preceding mid-term planning results).

If the latest planning effort provides the insights that major planning assumptions have changed, this knowledge can be used as iterative updates for the next revolving strategic or mid-term planning exercise (in this example, in 2021). Figure 4.3 illustrates the two-dimensional cascading flow – between plans/schedules on the same level, as well as between plans/schedules on different levels, but with the same planning period.

The next step in this revolving and cascading cycle is the first concrete flight schedule, designed around late summer/autumn with concrete resources

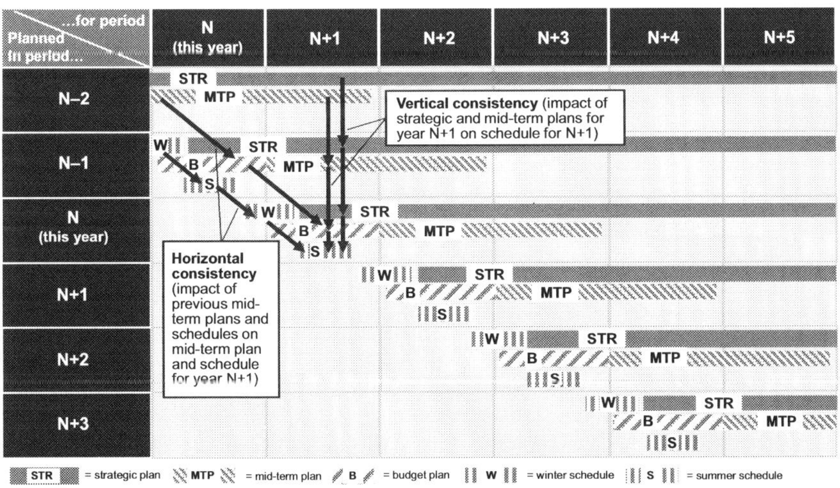

Figure 4.3 Schematic cascade of revolving planning processes and resulting plans/schedules

for a concrete schedule period (a few weeks up to a few months: see remark on scheduling frequency above). It is consistent with the latest mid-term plan within the three-year cycle described in the previous paragraph. In the concrete example referred to above, this first operational schedule would be prepared in August/September 2020 for summer 2021, and would represent the final update of the following mid-term plans:

- the last year of the plan designed in 2018 (for 2019 to 2021);
- the middle year of the plan designed in 2019 (for 2020 to 2022); and
- the first year of the current planning period in 2020 (for 2021 to 2023).

If assumptions need to be changed, the planners of course always use the latest information to do so, so the most recent plan always reflects the most up-to-date assumptions. As with the consistency between strategic and mid-term plans mentioned above, every concrete flight schedule stems from a 'pedigree' of preceding plans and schedules.

- *Vertical descent/consistency*: the flight schedule for the summer of year N+1 should be in line with the assumptions of all previous strategic and mid-term plans designed prior to year N+1 for N+1.
- *Horizontal descent/consistency*: the flight schedule for the summer of year N+1 is a descendant of preceding flight schedules (e.g., winter schedule N/N+1, summer schedule N, winter schedule N–1/N, etc.). If the respective goals and assumptions have not changed much, the latest schedule will be only an incremental improvement over the previous schedules. If there were disruptive changes in the market or regulatory environment, however, the latest schedule may look quite different from the previous ones.

4.4.11 Design of long-term abstract plans versus actual flight schedules

Strategic plans are usually less detailed than mid-term-plans, and mid-term plans are less detailed than actual flight schedules. Furthermore, strategic and mid-term plans cover a – theoretical – schedule period of 12 months in a full calendar year, while actual flight schedules contain realistic periods of six months or less (e.g., summer and winter schedules: see Figure 4.3). The transition between the high-level plans for a full calendar year and the actual flight schedules for a shorter period usually takes place in the course of the mid-term planning exercise; very often the first year of the mid-term plan (also referred to as a 'budget plan': see paragraph 'mid-term financial planning' below) is the first plan in the cascading planning process explained above to be designed for an actual schedule period, not for a calendar year.

Whereas the more high-level, annual plans are built on a few 'typical' traffic weeks, such as one high-demand week in September, one low-demand week in March and a holiday week in July, the actual flight schedules feature a much more granular view, determining demand and capacity for each traffic day. Schedules for typical traffic days may, but do not necessarily need to, recur. The latter phenomenon is likely to occur if the demand of an airline is highly volatile, for instance because of strong seasonality effects.

As an example, an airline could boost its demand with services to warm water destinations during the summer period, which may not be appropriate during the winter period. In this case, the horizontal descent concept illustrated in Figure 4.3 for actual flight schedules would have to be specified: every summer schedule would build on the preceding summer schedules, but not on the previous winter schedules – and vice versa.

Ultimately, for short-term planning or resource adjustments, the cascading logic is more trivial, since they always occur in the framework that has been defined by the most recent schedule.

4.4.12 Interplay between network planning and financial planning

It was stated earlier that the vast majority of all revenues and costs of an average airline are generated by its network. Consequently, there is a close connection between network planning and financial planning.

Long-term financial planning defines entrepreneurial objectives such as growth or profitability. These objectives need to be substantiated by the network – i.e., in the first step by the *long-term network planning*. The network scenarios or O&D demand bundles are designed in a way that the financial targets can be achieved – provided that the underlying assumptions on demand flows, price levels, regulatory environment and competition hold true.

The input of this planning step into financial planning is, of course, rather vague, since fares or unit cost based on O&D bundles – not on concrete flights or rotations – can be no more than rough estimates. Nonetheless, these estimates definitely provide valuable insights into incremental changes in terms of capacity, market shares and profit potential. Airlines will try to base investment decisions such as the purchase of aircraft on the results of long-term network scenario evaluation.

Mid-term financial planning deals with a more concrete evaluation of revenues and costs generated by the operated business model – i.e., by the capital and resources deployed for creating a certain value chain. There must be a balance between the allocation of resources, the respective financial results (such as cash flows or profits) and the financial or strategic targets. Mid-term planning commonly covers the next three years (cf. remarks on revolving planning cycles above), the most recent of which for the next year is also called the 'budget' or 'budget plan'.

At an airline, the allocation of resources for the mid-term financial plan is primarily determined by the network planning or scheduling process. The financial output of offered services and physical rotations is measured in expected revenues (assuming a certain level and mix of fares achievable in a certain market with a certain service), as well as in standard or planned costs (based on production KPIs such as CASK: see Chapter 3, section 3.1).

Although the underlying flight schedules for the two later years of the mid-term planning cycle are still somewhat theoretical and designed for one full year (based on so-called standard weeks), the most recent flight schedule already reflects a concrete flight schedule period (less than six months), checked for operational feasibility and including more realistic assumptions on demand flows, fare mix, unit cost, etc. In this regard, the accuracy of the budget (plan) is of course much better than that of the other two planning cycles.

4.4.13 Transition from sequential to overlapping process flow

Up to this point, the three network sub-processes described in this chapter have followed the strictly sequential flow illustrated in Figure 4.2. Around the beginning of the new millennium, however, this ideal sequence began to unravel, since the threshold between mid- and short-term planning is difficult to define in practice. Furthermore, airline demand has become even more short term and seasonal, to which airlines have responded by increasing the number of annual schedule cycles from some three schedule periods per year (e.g., summer, winter, holiday season) to ten or more.

Thus, most airlines have merged mid-term and short-term planning into one function, focusing instead on the continuous interplay between market planning and rotation planning. This new, more overlapping process flow is shown in Figure 4.4. As described earlier, in subsection 4.4.3, *market planning* assigns available seat capacity to expected demand per O&D, while *rotation planning* translates this rough capacity offering into physically viable roundtrips of aircraft and crews under operational restrictions.

Consequently, the horizontal split of processes between mid- and short-term planning has to some extent been replaced by a vertical split between market

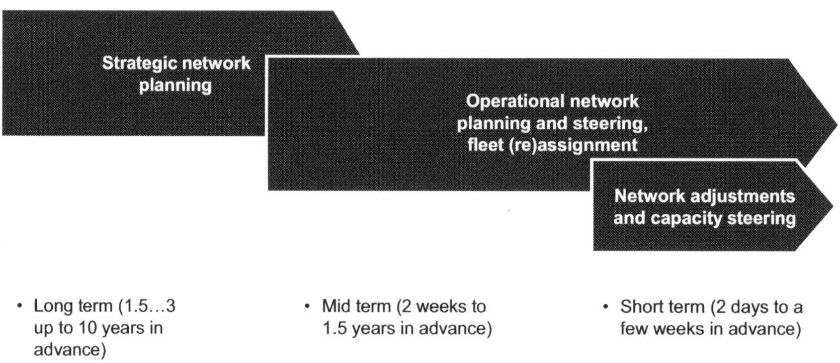

Figure 4.4 Overlapping process flow of network planning and steering[a]

[a] Franke, 'Network design strategies'.

and rotation planning. Furthermore, as described in subsection 4.3, there is also a split in the network management process flow of most airlines between long-haul and short-haul networks (and thus between the respective sub-fleets).

As is the case with many other managerial functions, the detailed design and timing of network management must observe the specific business environment of an airline, so aspects such as interfaces, decision rules and the timing of the handover between the process steps may vary widely. There are trends in the market, however, and the current trend prevalent for the past ten years is clearly an overlap between scheduling, short-term adjustments and deliberate resource reallocations.

4.4.14 Trade-off between stability and flexibility of plans

Unravelling network management processes along the timeline implies a compromise that an ideal, fully integrated planning process would not need: a trade-off between stability and flexibility. Since a stepwise process needs intermediate milestones and handovers, there is always the question of how detailed a planning result should be at a certain intersection or handover point.

One extreme perspective would be to strive for a very detailed plan at a very early stage. This would result in a concrete allocation of resources early in the process, avoiding a longer period with unclear allocation. The other extreme would be a paradigm with only a rough production scenario until shortly before the flight event, detailing the schedule and the respective resource allocation as late as possible.

The first scenario requires stability and suffers heavily from every disruption or change of business environment occurring after the schedule has been fixed. In this sense, the early finalisation of a schedule could be seen as artificial accuracy, ignoring the fact that there will always be external and internal changes, forcing the airline to adjust the schedule. The more adjustments need to be done, the more inefficient an accurate plan at an early stage will come across. The upside of this option, however, is the accurate starting point for every subsequent decision, which can be easily offered to the clients, and leaves the network manager with enough time to prepare the resources involved.

The second scenario acknowledges the fact that there usually are a lot of changes before a flight can finally depart, and that preparing an accurate schedule early in the process may be inefficient. A late scheduling process, on the other hand, requires an extremely flexible organisation, since it leaves the operational owners of the resources without a clear allocation notice until quite late in the process. The most severe flaw of this option, however, is the lack of a market plan that can be offered early enough to the potential clients. Without a detailed, published schedule, it is almost impossible to sell a single seat.

As usual, there is a sweet spot in the midst between two extreme scenarios: the overlapping process flow depicted in Figure 4.4 represents a pragmatic compromise, combining the timely definition of a marketable and operationally viable plan with the flexibility of adjusting the schedule if needed until quite late in the process.

4.5 Network management and adjacent processes

The value chain of designing, planning and steering networks is of course embedded into the overall process model of an airline. Since network management faces the unique challenge of considering commercial and operational aspects at the same time – and reconciling them in the most profitable way – this function features a wide range of interfaces with both commercial and operational functions. Figure 4.5 shows the interfaces between network management and operational planning processes along the timeline.

4.5.1 Fleet planning process

Most airlines start their resource planning cycle with their fleet, since aircraft require a certain lead time to be purchased or leased. Furthermore, aircraft represent the most significant capital lock-up of an airline, so it appears reasonable to plan their utilisation before all other resources (cf. network optimisation models in Chapter 5). Aircraft also need crews (cockpit and cabin) with the right licences for the aircraft type. For a single flight, the minimum number of crew members is defined by international standards (e.g., the European Aviation Safety Agency: EASA); a single airline is free to deploy more cabin staff to increase the service experience for its passengers, though.

For an entire year, the airline needs to employ multiple crews to operate one single aircraft to account for the highly regulated and limited labour time of airline crew members. This multiple is called 'crew complement' and varies with the type of aircraft, the labour laws in the respective country, individual employment contracts and the envisaged service level of an airline. A typical range for the crew complement of civil aircraft is one to 24, subject to aircraft type and size and dependent on flight distance and contractual provisions.

In the long run, the overall network strategy, containing so-called O&D demand bundles, needs to be aligned with the fleet strategy, ensuring a sufficient number of aircraft of certain types, providing the required capacity for the network scenarios. If aircraft are missing for a certain scenario, new ones need to be purchased (or leased) and commissioned in a timely manner.

For the medium term, the rotations translating the market plan into physical operations require aircraft units of a certain type plus suitable crews; this is ensured by the mid-term fleet and crew assignment plan. The planning of feasible crew rotations is also called 'pairing'. In the short run, the schedule must be filled with concrete crew resources ('rostering') and physical aircraft (tail number assignment). Short-term network adjustments, reflecting changes in demand, for instance, lead to respective adjustments in aircraft and crew assignment.

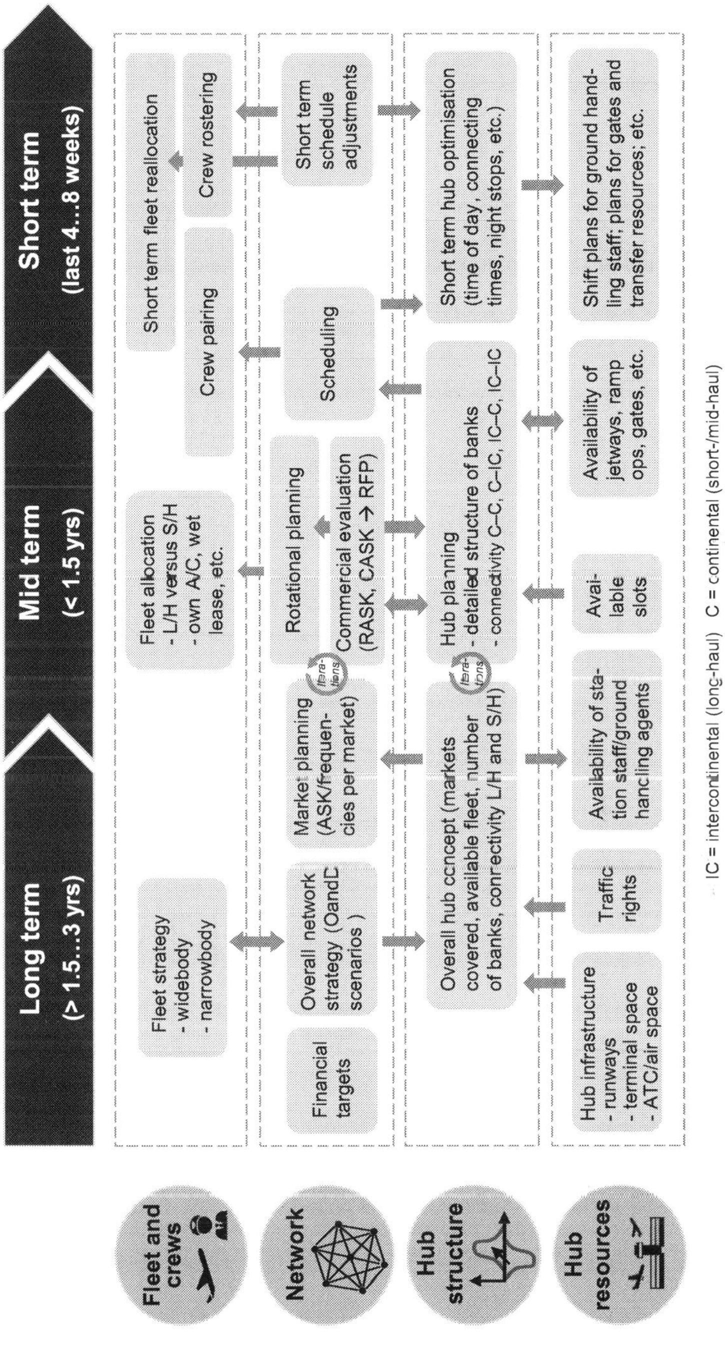

Figure 4.5 Interfaces between network management and adjacent operational planning processes[a]

[a] Franke, 'Network design strategies'.

4.5.2 Hub structure and hub resource planning

At hub and spoke carriers, network management has two more 'neighbours': a process designing and steering the hub structure (operational pattern: 'banks'), and another one planning the involved resources, such as slots in the air and on the ground, as well as resources handling aircraft, passengers and baggage. If an airline operates a hub, the hub needs to support the network concept (i.e., the schedule) in the most efficient way. The main performance drivers of a hub are as follows.

- *Connectivity*: the likelihood that a passenger arriving with an inbound (feeder) flight reaches his outbound flight within a reasonable timeframe (referred to as *minimum connecting time*, defined by the layout of the airport versus the maximum acceptable waiting time). Connectivity can be boosted by arranging inbound and outbound flights of the hub carrier in a peaky traffic pattern ('banks') – provided that a critical mass of flights is available.
- *Productivity*: connecting passengers from various inbound to various outbound flights boosts the revenue potential of the hub carriers but comes at the price of waiting times. Planned waiting times – together with unplanned delays due to airport congestion (causing holding patterns) or technical problems – massively reduce the productivity of all involved resources, especially of aircraft and crews. On the other hand, the bank structure can be optimised in a way that supports both connectivity *and* productivity. The latter is achievable if the rotational patterns served from the hub (distance and duration of the roundtrips) show a good fit with the gaps between the banks.
- *Market coverage*: a hub carrier aims to serve as many destinations as possible from its hub airport, since this increases the number of O&Ds it can offer, and thus the revenue potential it can generate. Furthermore, some hub carriers try to achieve an even spread of destinations in terms of geographic direction ('360 degree approach'), since an even spread also increases the likelihood of building O&Ds. This latter criterion is hard to influence, however, if an airport is, for instance, located in a peripheral geographic position relative to the mainland of a continent (e.g., Dublin or Lisbon in Europe).
- *Profitability*: a hub is, first of all, a cost centre, which, on the other hand, exerts some impact on the revenue generation of a hub and spoke carrier through the connectivity and service level it can sustain. A decisive driver of profitability is the balance between short-haul (very often feeder or de-feeder flights) and long-haul services. Long-haul flights are usually profitable, whereas feeder flights often lose money if they are calculated on a stand-alone basis. They facilitate indirect profits, however, since long-haul flights could not be operated without them; the cumulation of demand in the hub creates positive network effects. A hub that can handle only an under-critical number of long-haul destinations needs to fill its banks with

short-haul flights to achieve a critical mass overall. The more short-haul services that are offered, the more short-haul-to-short-haul connections with poor profitability will emerge – blocking the valuable space and time for the much more valuable short-haul-to-long-haul or even long-haul-to-long-haul connections

From the perspective of planning, the challenge for the interplay between network and hub planning is that airlines exclusively pursue a top-down approach: the network is designed and optimised first, and then the hub planners have to deal with whatever the network planners come up with – hub structure follows network structure. If, however, the specific conditions of a hub system require a change of the initial hub pattern (especially of the bank structure), airlines are mostly unable to calculate the bottom-up effect of this change on network and fleet performance (for more details on the interplay between network and hub planning, please refer to Chapter 6, section 6.5).

4.6 The 'transmission belt' concept

As distinct from other industries with a linear production model, and whose business models basically consist of sales-oriented and operational processes, airlines feature a third pillar: network management. Since the non-linear network of airlines is too complex and offers too many choices, the sales and operations functions alone are unable to identify the most profitable combination of revenue potential and capacity cost. This is the privilege of network management, since this function is the only one suited to combine the revenue and cost perspective of an airline. In a metaphoric sense, the network management function acts like a transmission belt between the marketing and sales process (demand perspective) and the production process (supply perspective), ultimately transforming available demand into sold tickets. This cyclical process is shown in Figure 4.6.

Market data are extracted in the form of demand forecasts per O&D (or O&D bundle) by a business intelligence function and translated into outlines of a market strategy. The latter defines passenger numbers and market shares per O&D or market, as well as ticket sales by region and sales channel. The market strategy is then transformed into network scenarios (featuring O&D demand bundles) and production concepts (fleet/asset strategy, hub concept). The network scenarios, together with the market strategy and production concept, must be suited to achieve the financial and strategic goals formulated by the airline.

In the medium term, the network scenarios are further developed into concrete schedules, which define a respective sales plan (per region and channel), as well as a suitable production plan (allocation of resources to routes/airports). In the short term, the schedules and plans need to be realised to unleash the envisaged profits. If the commercial or operational framework changes, network adjustments will be initiated, triggering also adjusted production plans (implemented by operations control) and in-field sales. The commercial

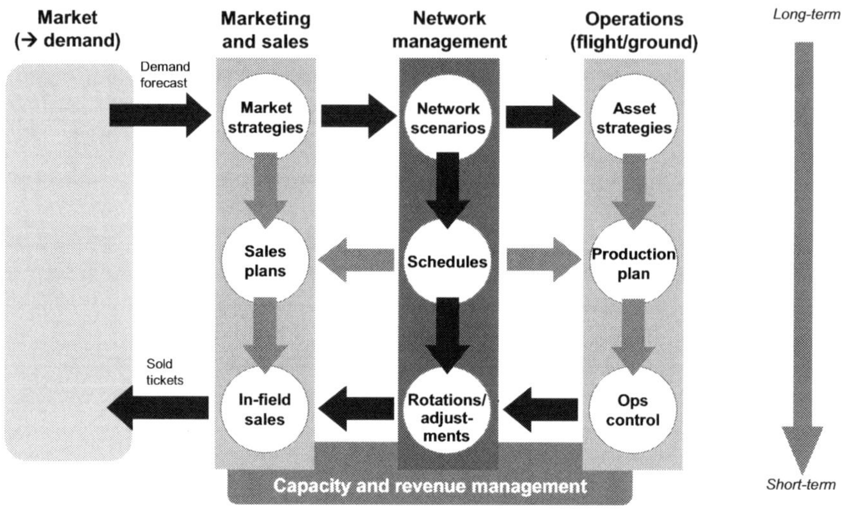

Figure 4.6 Transmission belt function of network design and planning at airlines[a]

[a] Franke, 'Strategic network management'.

optimisation of a flight is supported by a capacity control function called 'revenue management'. Besides operations control, revenue management is the most important short-term 'neighbour' of network management in the planning and steering value chain.

Revenue management has the task to sell a given capacity (e.g., number of seats or ASK produced by a flight) at maximum profit. Since the costs caused by operating the flight are fully determined at this late stage, 'maximum profit' is usually replaced by the objective 'maximum revenue/yield' (i.e., revenue per flight or RASK).

For further reading, a very comprehensive view on aviation network structures and state-of-the-art planning methodologies is, for instance, provided by Goedeking.[3] Deprosse and Händel contribute an instructive overview on network planning methodologies from the practitioner's perspective.[4]

Notes

1 Franke, 'Network design strategies'.
2 Bruce, P. J., *Airline Operations Control*, Abingdon: Routledge, 2020.
3 Goedeking, P., *Networks in Aviation: Strategies and Structures*, Berlin: Springer Verlag, 2010.
4 Deprosse, H., and Händel, M., 'Network planning and slot management', in Wald, A., Fay, C., and Gleich, R. (eds.), *Introduction to Aviation Management* (211–227), Münster: LIT Verlag, 2010.

5 Network optimisation models and planning tools

5.1 Linear programming for network capacity and resource allocation

Network planning problems at transportation companies are capacity and resource allocation problems. Demand (for seats on a certain O&D) needs to be matched with a suitable supply of capacity on routes suited to cover the O&D (point to point/non-stop or with one or more transfers). Since a network consists of a well-defined combination of edges and nodes, a transportation company such as an airline has a choice – at least in theory – to take very different itineraries through its network, and change the allocation of demand to capacity at every node for the subsequent edge(s). The complexity of the network, and thus the number of different options to move a passenger or shipment from one point (origin) to another point (destination) in the network, grows parabolically with the number of nodes, since *n* nodes can create a maximum of $\frac{1}{2}\, n \cdot (n \;\; 1)$ edges (i.e., possible routes connecting the nodes).

The number of options very quickly exceeds the capability of a human planner to evaluate the different options and find the most efficient solution for a certain transportation problem in the network. Problems of this kind are described and solved by operations research, a hybrid scientific domain blended from mathematics and econometrics. The standard OR approach for transportation (network) problems is linear optimisation or linear programming. The methodology was invented in 1939 by Russian mathematician Leonid Kantorovich, who solved a complex production problem with it, and finally received the Nobel Prize for his invention in 1975. The second problem solved by an LP methodology was already a transport problem (by US mathematician Frank L. Hitchcock in 1940).

The term 'programming' does not stand for computer coding in this context but is mathematical jargon for 'optimising'. Likewise, 'linear' does not mean that the mathematical problem is linear or in any way simple. Instead, the term refers to a core element of the LP methodology: the allocation problem is solved by a *linear* target function, combining a target value (e.g., maximum profit, minimum distance or minimum time) with a number of relevant constraints, represented by *linear* inequations. In a network problem, constraints could be the limited capacity of vehicles, limited speed or limited match between demand and offered service.

A specific subset of linear programming algorithms deals with optimisation problems featuring at least one variable that is not continuous, but discrete. For a transportation problem, this could, for instance, be the number of vehicles deployed on a certain route, since an airline can decide to use one aircraft or two, but no fraction in between. In mathematical terms, these variables belong to the sphere of *integer numbers* (0, 1, 2, 3, 4...).

An 'integer number' problem sounds like a less complex case than a 'real number' problem, since there are infinitely more real than integer numbers. For linear programming, however, the opposite is true: an integer (linear) programming task (IP) is much more complicated than a (real) linear programming task (LP). In real-life allocation problems such as in transportation, there is often a mix of integer and real number variables (e.g., the number of aircraft allocated to a certain route can be only an integer number, whereas price and time are continuous variables). This hybrid optimisation problem is consequently tackled by hybrid algorithms: the so-called mixed integer linear programming (MILP) solvers.

Although LP or MILP solvers such as CPLEX (an optimiser toolkit developed by IBM) are basically suited to finding efficient solutions for a transportation problem, modern network planning toolkits such as Netline (developed by Lufthansa Systems) contain much more than just LP, IP or MILP algorithms. The reason for this phenomenon is that some restrictions of transportation problems cannot be translated into the shape of a linear equation or inequation. For example, the economically most efficient allocation of capacity (e.g., aircraft) to demand (passengers) on a certain route depends not only on internal restrictions (e.g., available types of aircraft, available crew members, unit cost) but also, and heavily, on the behaviour of competitors on that route. Competitors can add capacity or undercut prices at any point in time, triggering overcapacity or price wars.

Likewise, competitors exert a strong influence on how the products or service levels of a carrier are perceived by its (potential) clients. A service level – in terms of product quality, number and timing of frequencies, etc. – may be perceived as absolutely sufficient by most passengers as long as only one carrier is offering flights on a certain route; as soon as a new entrant introduces a second offering in that market with a higher quality, clients will deem the offering of the incumbent less positive than beforehand. More details on capacity and resource allocation through linear programming can be found in Appendix A.

5.2 Logit models for the description of buying criteria under competition

In the unstable or moving scenarios described above, an initially and intrinsically *efficient* allocation of resources (for a single airline, especially for a monopolist) can very quickly turn into an *inefficient* allocation as soon as competitors change their offerings. The initial airline may decide to react to the move of its competitor(s), leading to an even fiercer competitive scenario, or to a new equilibrium (see further details in the case study in Chapter 8). In any event, the actions and reactions of competitors, especially those described by discrete values, cannot be translated into linear constraint functions.

For this complication of the transportation problem, mathematicians and OR experts have invented the so-called logit model approach. A logit model is designed to perform regression analyses for multiple, interdependent, *discrete* variables. Similarly to the phenomenon described above (LP versus IP algorithms), a logistical regression (for discrete variables) is much more complicated than a linear regression (for continuous variables). More details on the logit model approach can be found in Appendix A.

The application of the logit approach for our transportation/resource allocation problem *under competition* is the following: the first planning step for a network is usually the forecast of potential passengers on all relevant routes or O&Ds (i.e., expected demand flows). An airline that is a monopolist on a certain route could attract the entire demand for this route, unless it chooses a ticket price that deters potential passengers. As soon as a second or third carrier enters this market, the clients have a choice, and they will check various parameters to make their decision. The decision factors are usually a mix of hard factors (availability of certain O&Ds via that route, total duration of flight(s) on that O&D, non-stop versus connected service, number of frequencies offered per day/week, timing of the frequencies, product features included, etc.) and soft factors (perceived service level, reputation of carrier, bonus programme, involved partners, etc.).

Every single client could, of course, feature his individual decision behaviour based on these decision factors; a large number of potential clients usually show typical tendencies, though, with respect to their preferences for the above-mentioned hard and soft factors. If that is not sufficiently the case, the remedy could be to divide all clients into segments with more heterogeneous decision behaviour – e.g. business and leisure travellers.

In any event, the hard and soft decision factors mentioned above can be used as the describing variables in the logit model, since they ultimately trigger the decision of clients for a certain carrier on a certain route or O&D. Provided that the describing variables have been chosen in a realistic and consistent way, the logit function will deliver a solid prediction of the distribution of demand on a certain route/O&D to the carriers active in this market. Consequently, this application of a logit model is often called a *market model.*

5.3 Simplified example for planning with a logit model

For example, the logit model of *airline A* may predict an allocation of a certain forecast demand for a certain route (e.g., 100,000 passengers p.y.) to all three carriers active on this route in the following way:

- airline A: 40 per cent = 40,000 passengers
- airline B: 22 per cent = 22,000 passengers
- airline C: 38 per cent = 38,000 passengers

Airline A will start its planning exercise with an estimate of 40,000 passengers on that route – under certain assumptions for timing, frequency, service level and price. The assumptions (reflected in the describing variables and coefficients of the logit model) can of course be wrong, or competitors can easily spoil the result by deploying unforeseen changes to their offerings. For instance, the initial version of the logit model may assume that, if airline A offers its first flight on the respective route ten minutes earlier than airlines B and C, this will add 5 per cent to airline A's market share, since business travellers in particular have a strong preference for an early first flight. The reliability of this initial version of the model may be poor, however, for the following reasons:

- the descriptive variables of the market model are badly chosen – e.g., an important variable (= decision factor for clients) is missing;
- the coefficients of the market model are badly calibrated – e.g., the coefficients overestimate the importance of a ten-minute earlier first flight for business travellers; or
- the model is well designed and calibrated upfront, but one or more competitors still destroy the accuracy by adjusting their offering(s).

In the first two cases, airline A is forced to recalibrate its model, and make it more accurate. Due to the indirect correlations between variables, coefficients and final probability, this is a time-consuming process – which can be compromised at any point in time by the third case.

The accuracy of the market model predictions will be even worse if the relevant competitors use a significantly different market model, or ignore the results on purpose (e.g., to launch a price or capacity war). Any kind of diverging target function, or even irrational behaviour, heavily compromises the reliability of all involved market models. This is also true for poor or diverging input data. The latter effect became highly relevant when low-cost carriers attacked the established hub and spoke carriers in the late 1990s: to save money, LCCs avoided making use of the costly services of global distribution service providers such as Amadeus and Sabre – a behaviour that was totally contrary to the practices of traditional carriers.

5.4 Reliability of GDS transaction data

The flight ticket transaction data collected by GDS providers, and calibrated with the help of analytical service MIDT (Market Information Data Tapes; later acquired by Sabre), have been the high-quality basis for every demand forecast ever since the emergence of these providers in the early 1990s. With low-cost carriers staying outside the GDS inventories, the sample size of the MIDT demand data shrunk considerably, reducing the reliability of every single market model built on MIDT demand forecast data. On the other hand, the sample size is still larger than any other data source available in the market, so most hub and spoke airlines still use market models following the mechanism described above.

Even without the complications described in the previous paragraphs, a market model would not automatically deliver a perfect result for the efficient allocation of resources to expected demand under competition, since there are a multitude of financial and operational trade-offs that need to be evaluated before the final network scenario or schedule can be designed. For example, airlines usually own different types of aircraft, with different seat capacities and unit cost levels. Without any further restriction or competition, the best choice for an airline would be to deploy the largest aircraft available to be filled with a reasonable seat load factor (see Chapter 3: 'Trade-off between aircraft size and frequency'), since that option would minimise the unit costs for a given revenue, and thus extract the maximum profit. As already stated, however, in Chapter 3, section 3.2, this choice may still be suboptimal in a real-life environment with competition thanks to the S-curve effect.

If competitors offer a higher number of daily or weekly frequencies on a certain route, potential passengers may perceive that service level as superior, rewarding the respective carriers with a disproportionately higher passenger market share than might be expected from the capacity market share. In this scenario, it could be profitable to increase the number of frequencies as well, switching to smaller aircraft types to ensure reasonable seat load factors. The increased revenue through additional high-yield (e.g., business) travellers could then more than compensate for the less favourable unit cost position of the smaller aircraft. Trade-offs such as this one can be thoroughly addressed in a comprehensive, iterative network planning approach, as sketched in Figure 5.1.

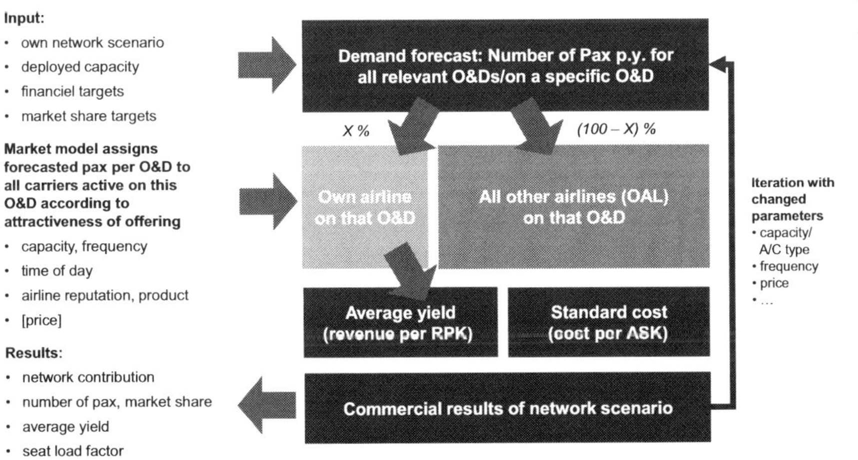

Figure 5.1 Concept of iterative demand forecasting and design of network and schedule[a]

[a] Franke, 'Strategic network management'.

5.5 Definition of strategic and financial targets

In the *first step* of this approach, an airline defines strategic and financial targets for its network (e.g., market shares in certain regions, growth, network contribution). Subsequently, a network scenario (later on: a concrete flight schedule) is designed, reflecting key parameters such as forecast demand flows per O&D, destination patterns, the available fleet, seat capacity, unit cost levels (e.g., CASK) and fare levels (e.g., RASK or RRPK). This network scenario is then evaluated by a market model (e.g., a logistic regression model, as explained earlier in this chapter). The market model compares the attractiveness of an airline's offering along various decision criteria with that of its competitors on the same O&D/ route, calculating the probability that passengers decide to buy a ticket of any of the carriers active in this market.

5.6 Calculation of market share

The result delivered by the market model in the *second step* is the market share of the carrier on all relevant routes and O&Ds, and ultimately the number of passengers in different booking classes, adjusted to the perceived quality and attractiveness of the offered services compared to competing offerings. In the *third step*, the passenger numbers can be translated into seat load factors per flight, as well as into revenue streams for this flight (measured, for instance, in RASK). The carrier decides for itself how granular the respective fare structure reflected in the market model should be; a simplistic structure (business class fare = '*normal tariff*' versus economy class fare = '*special tariff*') is less precise, however, though much more robust than a model that gets closer to the real-life fare structure with dozens of tariffs and sub-tariffs.

5.7 Estimate of revenues and costs of flight programme

In the *third step*, in parallel to the revenue calculation, the planned flight programme (network scenario or concrete flight schedule) is labelled with the respective operational costs (cf. the network contribution model introduced in Chapter 3, section 3.3), commonly with standard or planned cost per ASK (CASK). Revenue and cost calculation can then be combined in the *fourth step* to determine the potential commercial result of the chosen flight programme. This result, comprising network contributions/profits, market shares, yields and seat load factors, must be held against the initial financial and strategic assumptions, used as upfront input for the entire calculation model.

In the first iteration, it is more than likely that the outcome of the model (after step 4) does not match the input into step 1. On a detailed level, the following deviations may occur:

- the desired market share/passenger number is not achieved during the calculation;

- the desired seat load factors are not achieved; or
- the financial results stemming from the economic evaluation (steps 3 and 4) deviate from the initial strategic targets for the chosen network scenario and period.

5.8 Iteration of results

There are various levers to achieve a better match between input and output during the next iterations of the model. The adjustments are today performed by experienced professionals, understanding the various trade-offs and interdependences described in this chapter. In the future it is conceivable that there might be machine learning algorithms capable of supporting – if not replacing – human planners in this effort (see also Part III for future planning paradigms).

For example, the first iteration may end up with an unsatisfactory economic result, due to passenger numbers and revenue levels being below expectations on key routes. Since the deployed capacity remains the same, costs do not go down, compromising overall profitability. Network KPIs such as seat load factor may also assume critical values, indicating an inefficient scenario as well. Obviously, some of the initial assumptions need to be adjusted for the second run of the model to get closer to the desired result. If passenger numbers and – consequently – revenues do not match the plan, there are various levers to counteract the shortcoming:

- increase the number of frequencies on critical routes/O&Ds to benefit from the S-curve effect (cf. Chapter 6, section 6.3);
- improve the timing of flights on critical routes/O&Ds to create a better match with the requirements of clients, especially those of demanding customers such as business travellers;
- improve product quality (e.g., seat space/pitch, amenities, in-flight entertainment, catering);
- invest money into brand building (long-term) and promotions/marketing campaigns (short-term); or
- reduce fares.

These levers may increase the passenger numbers, provided that competitors do not take similar measures and thus neutralise the envisaged effect. If the deployed market model takes these decision criteria into account as well – as variables of the logit model described earlier in this chapter – it will reward the respective improvements by assigning more passengers to the routes. These changes come at a price, though.

An increased number of flights on a route, for instance, may raise enough additional demand to fill the incremental capacity with the same seat load factor as before – but this would be mere coincidence, if not a best-case scenario. In a less fortunate scenario, the passenger number does go up, but not proportionately with the added capacity, thus compromising the overall seat load factor.

This scenario could become counterproductive, since the cost of the unused capacity (i.e., CASK) would eat up the increased revenue. Ultimately, the number of passengers would go up, but profitability would go down. The remedy for this downside is the deployment of smaller aircraft with less seat capacity. Through this move, the number of frequencies could be increased without worsening the seat load factor. But even this second move is not free of risk: smaller aircraft tend to feature higher unit cost (*ceteris paribus*, same technological generation), so, again, it is not trivial to transform more passengers into higher profits.

5.9 Impact of timing on the perceived attractiveness of an airline's offering

An improved timing initially is a free-of-cost lever, but it is by no means a trivial one. First, deterministic market models tend to overrate the positive effect of outpacing competitors. For instance, a typical market model would come to the conclusion that departing ten minutes earlier in the morning is suited to alienate a significant amount of time-sensitive clients, especially business travellers, from competitors. As long as GDSs use the same assumption, the effect will really occur as a self-fulfilling prophecy, since clients accept the underlying assumption and follow the resulting recommendation. The real effect of a ten-minute discrepancy could be much smaller or close to zero in real life, though, turning the lever into a purely theoretical one, featuring artificial accuracy.

Even if the inherent assumption is realistic, and time-sensitive travellers do indeed reward an increment in departure time as small as ten minutes, nobody can stop the competitors on that route from coming to the same conclusion and copying the initial move (contingent, however, on slot availability at that time). This sequence of actions would very quickly end up in a stalemate scenario, making the timing of flights useless as a lever for increasing passenger numbers.

In the worst case, all competitors would use market models with identical assumptions and algorithms, generating exactly the same recommendations, and removing every differentiation after a short while. And, ultimately, departing ten minutes earlier than the competitors may simply be impossible if the respective slot is not available at the outbound or inbound airports.

5.10 Impact of product quality on the perceived attractiveness of an airline's offering

Investing in product quality leads, as opposed to the previous lever, to a physical and thus sustainable differentiation; the required investments may nonetheless be significant. Furthermore, improving product features, or even equipment, is probably not viable at short notice, so not relevant for the iterative cycle described in this chapter as far as short- and mid-term schedules are concerned. The same is true for investments into brand reputation and marketing campaigns. For long-term, strategic network scenarios, however, all above-mentioned levers for increased passenger numbers can be taken into consideration.

Consequently, increasing the number of frequencies – potentially with smaller aircraft – and adjusting the timing of flights on a certain route are the prevailing levers to influence the revenue side of business. For more strategic planning, levers such as investments into capacity, product features and brand reputation are also important.

5.11 Impact of pricing on the perceived attractiveness of an airline's offering

Pricing is always a special lever. As stated earlier, price is usually not part of the logit or market model but is considered separately. Using promotional pricing, namely discounted fares, to attract more demand must be considered very thoroughly, and exerted with utmost care. First, a lowered price level has the unpleasant tendency to stay forever in a market, since competitors may feel forced to match discounts, and clients take the new, lower price level for granted.

Second, discounts may attract more clients, but they do not guarantee higher revenues – depending ultimately on the price elasticity of demand in the respective market. If potential clients react with strong price elasticity, because they are very sensitive, then the additional volume will make up for the reduced price. If, instead, demand is rather inelastic, the overall revenue will be lower than before despite the increased volume. This correlation is particularly true in monopolistic markets (and can be used to calculate the ideal price for a monopolist), but also applies in oligopolistic markets.

To make the latter case even more complicated, there is not a single price elasticity of demand per market, but many; in reality, a matrix with cross-elasticities between all competitors can be observed and calculated. For example, if *carrier A* lowers its price by 10 per cent in a selected market with three players, it may experience a 5 per cent stimulation effect from former non-travellers, a 7 per cent enticement from *carrier B* and only a 2 per cent enticement from *carrier C*. The reason for these differing attraction rates may lie in the initial relative price level between the three competitors, but also in different reputations and quality perceptions. In this respect, the cross-elasticity follows a mechanism similar to that described for the logit model earlier in this chapter.

In essence, all five levers listed in section 5.8 may be helpful to raise passenger numbers and move closer to a match with the initial targets; because of the complexity of the trade-offs, however, the outcome may also be the opposite, or come at a price that is not affordable. It is thus essential that every iteration step is closely monitored with respect to its convergence – i.e., its tendency to reduce the gap between output and input in terms of performance indicators. If the gap gets bigger again in a certain step, the chosen measures were counterproductive in achieving the initial targets. The iteration process can be finished once a predefined threshold for the gap between output and input is reached (e.g., 1 per cent). Every further step would then create minimum convergence with considerable effort.

5.12 Iteration process in case of an unfavourable cost position

Likewise, the iteration process could start with a situation in which the passenger numbers already show a good match with the expected figures (used as input into the first iteration), but the economic result is nonetheless poor, since the costs are higher than planned. The reasons for this effect could, for instance, be a higher share of smaller aircraft with higher unit costs, or more complex rotations of aircraft and crews, causing overtime payments for the crews and overnight parking fees for the aircraft. In this case, the input for the next iteration step needs to be adjusted in a way that reduces cost, for instance by deploying more efficient rotational patterns, or larger aircraft with a superior unit cost position.

Immanent countereffects need to be kept small, however, since the next iteration may otherwise lead further away from the targets instead of converging. For example, deploying larger aircraft on a certain route may require a drop in frequencies to ensure satisfactory seat load factors by keeping the overall capacity offering in that market constant. Nevertheless, a reduced frequency pattern may be perceived as inferior to other airlines by time-sensitive (business) travellers, ending up in a 'penalty' for this carrier in the form of passenger losses. In this case, the lost revenue may consume the saved costs through scale effects.

5.13 Convergence of market model

Ultimately, it is a purely technical question how many iterations are needed until the entire mathematical model (market model plus commercial evaluation) achieves a satisfactory level of convergence. Usually, an appropriate threshold will be defined before starting the iteration process. If the algorithm then continuously converges – i.e., produces dropping discrepancies between output and input parameters in every single iteration step – it is very likely that the threshold will be reached after a finite number of iterations. The parameter values chosen for the final iteration step are then accepted as solutions for the iterative problem.

If the model does not converge at all, however, or not sufficiently to undercut the predefined threshold, the entire mathematical model may not be solvable. In this case, the constraints of the involved equations may be too rigid, or the financial targets simply too optimistic. Other levers may be available in theory, but not at short notice – e.g., replacing an aircraft with relatively high direct operating costs (DOC) by a more efficient aircraft.

6 Impact of network design on airline planning

6.1 Network design and planning scenarios

An airline network is a highly dynamic structure used as the core production platform to physically transport passengers and/or cargo. It is constituted by the combination of routes and all resources required to perform the transport service on these routes. On the top level, the complexity of designing and planning a network is determined by its sheer size. The more destinations that are included, the more routes that need to be flown, and the more resources (aircraft, crews, etc.) that are required. The number of viable scenarios to combine these resources grows disproportionately, if not exponentially, with the number of routes and resources involved.[1]

For hub and spoke networks, this effect is stronger than for P2P networks, since a hub establishes transfer opportunities, and thus additional options to connect destinations with given resources. Whereas, for example, a P2P network with ten destinations can easily be designed and planned in an Excel sheet, a hub and spoke network with, say, 100 destinations worldwide requires a variety of sophisticated optimisation models, as well as significant computing power.

6.2 Network and market coverage

6.2.1 Basic approach to market coverage

Demand for air transport occurs between city pairs – i.e., between origins and destinations. Origins and destinations represent local markets, featuring specific customer segments and preferences. The primary goal of designing an airline network is to capture as much of this demand for air travel as possible in a sustainable way. This implies that growing a network is always desirable as long as it can be served profitably, since a large number of destinations increases the theoretical revenue potential and fosters scale effects. On the other hand, it increases the number of routes to be offered, as well as the investment risk. If we assume that a network will always start with the most profitable and least vulnerable routes, every incremental destinations or frequency adds a disproportionately higher risk than the preceding on. Thus, the vital question for every airline is *how many* and *which* local markets should be covered by the network.

The immanent scale risk described above is also dependent on the chosen business model. Whereas low-cost carriers can opt for a 'cherry-picking' approach and select profitable routes only, network carriers must deal with both profitable and less profitable routes, since their hub rationale causes highly complex and interdependent network patterns. They tend to cross-subsidise feeder flights with trunk routes, as well as (less profitable) connecting flights with (usually profitable) local demand, without knowing exactly how big the cross-subsidisation will ultimately be.

It becomes obvious that hub and spoke carriers generate a significant dilemma for themselves by establishing a hub to capture as many traffic flows as possible. To run the hub efficiently, they need good utilisation of the airport's capacity – i.e., a large number of connections. To make this system attractive for both local and connecting travellers, they are bound to offer a broad range of destinations to all passengers, who could otherwise use a competing hub or direct flights. As a result of this fundamental need for scale and market presence, hub airlines must achieve a more or less comprehensive coverage of major markets within their reach, a goal known as the '360 degree model'.

The result is, as stated above, a risky mix of profitable and less profitable routes, with sometimes intensive cross-subsidisation between them. Moreover, a significant share of resources, as well as of management capacities, is usually absorbed by the least profitable connections facilitated in a certain hub, such as domestic or intracontinental connections. This leaves little room for manoeuvre for the highly profitable long-haul connections. On the other hand, understanding this effect means being able to fix it (for details, see section 6.5).[2]

6.2.2 360 degree approach in hub airports

In general, the 360 degree approach provides a sound basis for efficient hub operations, since it helps achieve a favourable balance between different market types and traffic directions. Due to the effects of global time zones, long-haul westbound traffic has preferred departure times in the morning till noon at almost every airport worldwide, because passengers then conveniently reach their destinations in the afternoon or evening of that same day. North- or southbound flights hardly leave the time zone of the originating airport, so departures throughout the day are viable, depending on how distant the destination is. For eastbound flights, the anti-clockwise rotation of the Earth has a retarding effect, so airlines are forced to offer night flights, departing between the afternoon and midnight of the previous day. Airports covering a good mix of all these directions are able to achieve a very dense and homogeneous bank structure.

Nonetheless, there are also hub airports that, due to their geographic position, can never support a 360 degree approach, especially airports in geographic rim positions:

- very far north (Northern Hemisphere) or south (Southern Hemisphere) – e.g., Toronto, Stockholm or Cape Town
- at the western or eastern rim of a continent – e.g., New York, Lisbon or Beijing
- very far from major markets in the Northern Hemisphere – e.g., Sydney or Santiago de Chile

The resulting directionality of traffic leads to larger gaps in the bank structure of the respective hub. On the other hand, size is a good remedy for this structural flaw, since it fosters continuous hubbing (e.g., in New York, JFK).

Likewise, some airlines were forced to downscale their hubs, such as Lufthansa in Munich, compromising on the 360 degree approach. This move bears the risk of entering a downward spiral, whereby reduced market coverage results in declining attractiveness, which then dilutes demand and finally competitiveness in terms of scale and utilisation.[3] Real-life examples prove that this challenge can be handled, though, and that a consolidated hub can be more sustainable than an oversized one (for details, see section 6.5).

6.2.3 Effect of airline alliances

Another very powerful means to increase market coverage is partnering with other airlines, either on a bilateral level (e.g., interlining, code-sharing or joint venture) or by joining an alliance such as Star Alliance, oneworld or SkyTeam. International airline alliances hit their peak between 1997 (when Star Alliance was founded) and around 2005, because they created larger networks with seamless travel processes for their clients, and achieved scale and market power for higher efficiency. After 2005 their strategic importance stagnated, since ongoing deregulation made airline mergers easier than before; the negative trade-off between additional scale and additional management complexity limited the further expansion of alliances.

Furthermore, the strongly growing low-cost carriers, as well as most of the emerging Middle East carriers, refrained from entering alliances. The big three alliances still exist, however, and they still ensure global reach and convenience for many loyal clients beyond the home turf of the single member airlines. Whereas the coordination of some crucial processes or policies is forbidden by international anti-cartel laws (especially in pricing and revenue management), the alignment of network schedules is one of the major benefits of alliances.

6.2.4 Ramping up a network

For established carriers, building their network is a continuous task they deal with every single day. Thus, accommodating growth usually means incremental changes of the network – a new destination, a larger aircraft type or additional frequencies on high-volume routes. In some cases, however, changes can be more than incremental:

- if a carrier is forced to reduce or even close down its business, for instance in a major crisis such as the 2020 COVID-19 pandemic, and then ramps its business up again; or
- if a carrier embarks on a strategic expansion path.

In both cases, the respective carrier needs to achieve a balance between the capacity expansion, on the one hand – creating additional operating costs – and the increased revenue potential, on the other. Although the additional cost of any expansion becomes effective as soon as the new capacity is in place, the expected revenue is uncertain. This is critical, since demand is even more unforeseeable after a crisis than in a stable business environment.

Luckily, capacity expansion can be achieved in different ways, which gives the carrier a certain leeway to mitigate the capacity ramp-up risk: contingent on the geographic location of the hub and on the fleet composition, an airline can choose between different route types, aircraft sizes and service frequencies when it expands its network. As for the entire network, there is so single solution for the optimum expansion path; the following rules of thumb for a balanced expansion have been successfully executed by various hub and spoke carriers, however:

(1) the ramp-up of routes to major continental destinations (trunk routes or markets with high-value connected traffic potential), including hubs of competitors;
(2) the launch of long-haul flights to primary intercontinental destinations (long-haul trunk routes), including hubs of competitors;
(3) the addition of secondary continental routes;
(4) the extension of the long-haul network to secondary destinations (especially non-hub bases of competitors);
(5) the establishing of services to small continental routes (provided that a dominant or even monopolistic position can be achieved in that market);
(6) if further growth occurs, the addition of frequencies to destinations under points 1 to 4 above, potentially combined with a switch to slightly smaller aircraft types on routes to secondary destinations; and
(7) if further growth occurs, the switching (back) from smaller to larger aircraft types to destinations under points 1 to 4 above, in order to benefit from the decrease in unit cost and subsequent improvement of competitiveness.

Steps 1 and 2 mainly aim at quickly achieving a solid base load for the expanded or reactivated network at limited risk, whereas steps 3 and 4 can be a direct attack on the home carrier of a target country, trying to entice O&D traffic away from that home carrier in its non-hub bases.

6.2.5 Extraction strategy

For example, a US hub carrier could try to alienate Lufthansa clients from large non-hub markets such as Düsseldorf, Berlin or Hamburg by offering non-stop flights to major destinations in the United States – e.g., to New York, Chicago or Atlanta. This will be seductive for many German travellers if Lufthansa does not offer direct flights on these O&Ds but only as connected flights via Frankfurt. Furthermore, some of the alienated German clients may be interested in connecting via their primary destination in the United States to further destinations in North, Central or South America. It is very difficult for the attacked home carrier to defend against this passenger drain if the non-stop services are compelling enough to be perceived as viable alternative to the connected flights of the home carrier. A foreign (inbound) carrier usually has a weaker reputation in that market than the home carrier, though, so is usually forced to grant discounted fares.

Other examples for this *extraction strategy* are the networks of the major Gulf carriers (Emirates, Etihad, Qatar Airways), as well as of Turkish Airlines. Emirates, for instance, has over the past 15 to 20 years systematically built up services to and from large *hubs* in Germany, the United Kingdom, France and Italy (e.g., London Heathrow, Frankfurt Rhein-Main, Paris Charles de Gaulle, Milan Malpensa), but subsequently also to large *non-hub cities* such as Düsseldorf, Lyon, Manchester and Hamburg. In the first category of destinations, competition is of course fierce, since Emirates attacks the national flag carriers in their hub. In the second category of destinations, however, the national flag carriers have difficulties stopping Emirates from alienating their local clientele, since they need to fill their hubs, and cannot offer many direct flights from non-hub cities to Dubai.

Furthermore, the P2P demand from European non-hub cities to Dubai (as well as to other hubs in the Gulf region, such as Abu Dhabi, Doha and Bahrain) is limited; the larger benefit for Emirates lies in transferring these European clients to destinations further east, such as Greater China, Southeast Asia and Australasia. Thus, the real competition through Emirates' extraction strategy occurs on *eastbound long-haul O&Ds*, not that much on *inbound traffic to Dubai*. Or, from the perspective of, for instance, Lufthansa: losing German clients on the mid-haul O&D Düsseldorf – Dubai is bad, but not unusual, since Emirates as the hub carrier in Dubai is in a stronger position than Lufthansa, the latter serving both Düsseldorf and Dubai as a non-hub carrier. It is more painful, however, to lose high-value passengers on long-haul O&Ds such as Düsseldorf – Singapore (connected via Frankfurt or Munich with Lufthansa, via Dubai with Emirates).

The same is valid for the extraction of transfer traffic by Emirates from European hubs; it would be highly unfortunate for Air France, for example, to lose formerly loyal passengers on the O&D New York – Hong Kong (connected via Paris Charles de Gaulle, CDG) to Emirates, provided that Emirates is able to persuade the respective travellers from New York to continue their itinerary in Paris with Emirates (transferring them to Hong Kong via Dubai) instead of with Air France. Luckily for the attacked hub carriers, this scenario is much less probable than the previous one, since the clients from New York would have to switch from a one-stop flight to a two-stop flight – a loss of convenience that most passengers would either avoid (especially business travellers) or accept only for a significant discount.

For the previous case (extracting passengers from European non-hub cities and flying them to a hub in the Gulf region or beyond), the main arguments of the Gulf carriers are:

- direct flight to the Gulf region, and a good connection to East African, Chinese, Southeast Asian and Australasian destinations
- attractive service level in terms of timing and number of frequencies
- superior value-for-money ratio (mostly through discounted fares)

6.2.6 Examples of airlines ramping up their networks

Strongly growing airlines from emerging regions such as Turkish Airlines or the Gulf carriers have diligently pursued the seven-step strategic path sketched above. In particular, Turkish Airlines has been very successful over the past decade by building a domestic, short- and mid-haul feeder network (in Europe, North and East Africa, the former Soviet Union and the Middle East region), upgrading its fleet, product concept and hub performance and, finally, expanding its intercontinental network. Once a certain destination works well and the demand grows, the airline adds a second or third frequency per day, to make its offering on that route even more compelling and become the dominant carrier in the respective market.

Starting in the early 1990s, Emirates has followed the same strategic path, followed later on by Etihad and Qatar Airways. The main discrepancy between the Gulf carriers and Turkish Airlines is that the domestic market in Turkey is large, while it is tiny in the Gulf countries (except Saudi Arabia). Thus, the Gulf carriers need to replace domestic and short-haul traffic by more mid-haul feed, such as from East Africa, Egypt, Central Europe and India. Consequently, the fleets of these carriers contain only a few narrow-body aircraft (Etihad, Qatar Airways), or none at all (Emirates).

This shortage of domestic and short-haul feeder traffic is a significant limitation for the growth of the Gulf carriers. They may be very successful in the high-value long-haul business (sixth freedom traffic: see explanation in section 6.6), but this segment accounts for only 10 to 15 per cent of worldwide air traffic. In addition, they strongly compete for the long-haul traffic with their hubs located in very close vicinity to each other, plus aggressively growing carriers from adjacent regions such as Turkish Airlines that extract their clients from exactly the same mid-haul markets.

6.2.7 Risks of ramp-up strategy

These limitations also hint at the fundamental risk imposed by the described growth strategy: it costs a fortune and shows mediocre resilience to intensive competition or disruptions such as a pandemic or financial crisis. Filling the first mid-haul flight from a major non-hub destination is often easy, especially with an attractive service concept and discounted fares. To stay competitive and attract time-sensitive business travellers, however, a second or even third daily frequency soon needs to be added, which requires more and more cross-subsidies from the long-haul flights.

Either the additional flights worsen the overall seat load factor or the carrier damps that effect with smaller aircraft types but pays the price with a deterioration of the unit cost position. As soon as economic stagnation, or even a crisis, occurs, these routes consume more cash than the long-haul flights can contribute, leaving the respective carrier with an unsustainable position.

6.3 Network and fleet structure

The most important resources needed to operate an aviation network are aircraft, slots and crews. In addition, ground handling and aircraft maintenance capacities are required. Aircraft (if owned) lock up huge amounts of capital, so establishing the right fleet structure is immensely success-critical for an airline.[4]

6.3.1 Impact of business model

The most efficient fleet structure is highly dependent on the business model chosen by an airline. For low-cost carriers operating P2P networks mainly on a continental or domestic level, it is feasible to operate one or two (narrow-body) aircraft types only. In contrast, hub and spoke carriers, with their complex mix of short- and long-haul services, as well as of flights with a focus on P2P versus connected traffic, are forced to operate more heterogeneous fleets, with four, five or even more different aircraft types.

Fleet commonality improves the economics of a network, since it keeps the complexity of maintenance and crew assignment as small as possible. As stated above, hub and spoke carriers are forced to continuously evaluate the trade-off between fleet commonality and a larger number of aircraft types, the latter enabling the airline to accommodate the specifics of different markets (in terms of range and capacity). In general, the basic rule of thumb is that an airline should always deploy the largest aircraft it can fill in a certain market, since unit costs (i.e., CASK) decrease with growing aircraft capacity. There are at least two limitations to this rule, though.[5]

- A larger aircraft tends to have a better unit cost position (and, thus, incremental profit) than a smaller one. This does not mean, however, that the largest available aircraft ensures the highest *total profit* on a certain route. Depending on market as well as aircraft characteristics, it may happen that the gap (i.e., profit) between the curve of the accumulated yield and the total cost of available aircraft is not largest for the maximum number of passengers and the largest aircraft, but for a smaller aircraft with fewer passengers. For most airline managers, this rationale is severely counterintuitive, since they were trained in the paradigm of unit cost and scale effects. The above-mentioned exception from the rule means that an airline could decide to leave some of the potential passengers on the ground, with parts of the theoretical revenue potential untapped.

- In certain highly competitive markets, it may be wiser to cover the available number of passengers with smaller aircraft and, in return, with a higher number of services per day and week ('frequency'), compared to the scenario with the largest available aircraft for that route. This scenario is driven by the so-called 'S-curve effect', and will be explained in the next paragraph.

6.3.2 The S-curve effect

As explained in the previous paragraph, it may be more efficient in a highly competitive market to switch from X flights per day with the largest available aircraft type to X+1 or X+2 flights with a smaller aircraft type – despite the fact that the larger aircraft type ensures a better unit cost position than the smaller type. Such a decision would acknowledge the so-called 'S-curve effect', an empirical effect that occurs in the following manner:

> If a carrier enters a market with limited capacity (available seats x frequency), it will, first of all, have a niche position. Any frequency extension from that low level will not be fully 'rewarded' in the beginning by the passengers, since their perception is blurred by the more comprehensive offering of incumbents on this route. Thus, an increase in frequency share will only create a *disproportionately lower increase* in passenger share.[6]

Vice versa, a carrier holding a strong position in a market and offering a significant number of seats and frequencies will benefit from the opposite effect, operating in the supercritical or dominance area of the S-curve. Thus, a *further rise* of an already strong frequency share will entail a *disproportionately higher increase* in passenger share, until, ultimately, the saturation area is reached. For example, a niche player could add 10 per cent more capacity but attract only 5 per cent more passengers (i.e., less than a 'fair' share), while an incumbent may achieve a 15 per cent passenger increase with the same incremental capacity addition (i.e., more than a 'fair' share).

The S-curve phenomenon, which is, incidentally, independent of any pricing effect, is illustrated in Figure 6.1. The impact of pricing, which of course strongly affects customer preferences as well, will be discussed in the case study presented in Chapter 8.

The conclusion from the S-curve phenomenon is that, if an airline is a niche player on a certain route, and does not intend to leave the niche, it can deploy the largest aircraft type it can fill in this market, and optimise its unit cost position. On the other hand, a carrier with a strong to dominant position in a certain market may decide to compromise on aircraft size and unit cost, and invest in more frequencies instead; this would help fully leverage its dominance on the revenue side.

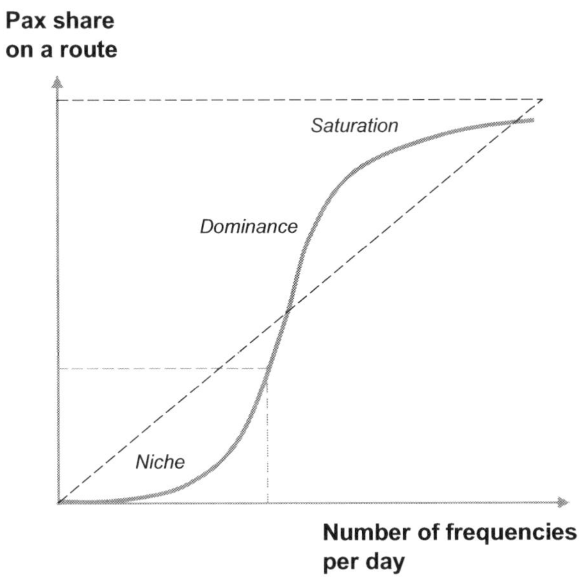

Figure 6.1 'S-curve effect' in airline markets[a]

[a] Franke, 'Network design strategies'.

In practice, a niche position usually occurs below 10 per cent market share, whereas a dominant position can be established with market shares above 35 per cent. These thresholds are no more than rules of thumb, however, which may vary with the concrete competitive environment on a certain route or O&D. For example, it has quite an impact on the thresholds whether two, three, four or even more airlines compete in that market. In any event, there is a no man's land between the niche and incumbent positions (in the example above, between 10 and 35 per cent market share), which usually proves to be unsustainable in the long run. A market share of, say, 20 per cent is too weak to compete with an incumbent, and too costly compared to a niche player.

6.4 Network and human resources

Human resources are – besides aircraft and airport slots – the third indispensable element of aviation networks. This resource cluster mainly comprises crews (for cockpit and cabin) and ground handling staff (for aircraft, passenger and baggage handling). The interplay between aircraft and crew is quite complex, since operating an aircraft has fewer restrictions than deploying crews. Crew members need to own the right licence to work on a certain aircraft type ('type rating'), and their working hours are restricted by law and by contracts.

Consequently, the roundtrips of crew members ('rotations') need to vary from roundtrips of aircraft. Planning the roundtrips for the crews is a challenging task, since the schedules must not only ensure a seamless staffing of the aircraft but also observe legal restrictions and the personal availabilities and preferences of crew members.[7]

This complex task is usually tackled with the help of a two-step approach: first, the roundtrips of aircraft are coupled to potential roundtrips of crews with the required rating for that aircraft type ('pairing'), until all aircraft are perfectly staffed throughout their entire operational schedule. Pairing is usually performed two to three months prior to a scheduled flight, and does not yet include the individual names of crew members or the tail numbers of aircraft. These detailed pieces of information are added in a second step ('rostering'), when the paired rotations are translated into concrete working plans for individual crew members and aircraft, commonly one month in advance.

Since the motivation of crew members is key for a flawless customer experience, many airlines grant a certain flexibility and involvement to their crews with respect to rostering. For example, crew members are often entitled to define some of their destinations themselves ('self-assignment'), to fly together with certain other crew members ('co-pairing') or to swap duty days with other crew members. These concessions add some complexity to the planning process, but usually more than pay off in terms of improved crew satisfaction and performance.

Ultimately, on the day of operations, unexpected irregularities related to the crew such as illness need to be handled at short notice, which is one of the dimensions handled by a disruption management/recovery function. Disruption management is a task taken care of by the operations control centre (OCC) of an airline. The handover between crew planning and the OCC usually takes place 24 to 48 hours prior to the day of operations.

6.5 Network and hub structures

Hub and spoke carriers operate one or more hubs to create beneficial scale effects. By accumulating demand in one spot, they can serve more O&Ds, capture more revenue potentials and deploy larger aircraft than carriers with a focus on P2P services. As mentioned before, the downside of this business model is the tendency towards complex operations and less productive assets, caused by the inherent interdependence of resources such as the waiting time between inbound and outbound flights. In addition, operational stability is more critical than in network structures without a hub, since peaked traffic structures ('banks') increase the risk of congestion and delays.

A hub is, first of all, a production platform, and then, second, an airport with a local market. Its major success factor is its ability to capture as many traffic flows (and, thus, revenue potentials) as possible, at minimum cost. On a more detailed level, its major KPIs are *connectivity* (its ability to connect inbound and outbound passengers in a convenient way), *productivity* (first of aircraft, then of all other involved resources), operational stability and the profitability of offered connections.[8]

The dominant driver of a hub is its sheer size and economic potential, determined by the strength of its catchment (i.e., number of people living close to the hub) and the wealth and mobility level of potential clients within this catchment. In fact, aviation – like many other industries – is a scale business, so size matters. Another crucial performance parameter of a hub is the balance between local demand and connecting traffic; as mentioned before, direct services often generate higher ticket fares than transfer connections, so carriers tend to cross-subsidise transfer traffic with local traffic, as well as feeder traffic with long-haul services. Airlines operating a hub with a limited local demand are more vulnerable to competition and price volatility than those with a strong local traffic share.

6.5.1 Bank structure of a hub

In the never-ending search for more efficient operations, hub and spoke airlines developed around 1980 the paradigm of compressing their inbound and outbound flights into peaks, increasing the probability that an inbound passenger can reach an appropriate outbound flight to his final destination without too much waiting time. This concept boosts the achievable connectivity in a hub by increasing the number of potential outbound flights reachable for an inbound flight within a predefined waiting time (usually called a 'hit'). A 'qualified hit' is a connection that not only works in theory but is demanded by the market – i.e., passengers really use this connection. The higher the number of qualified 'hits' the higher the share of the theoretical traffic that the respective carrier can cover by accumulating demand in its hub.[9]

To extract the full effect from this paradigm, hub and spoke carriers need to continuously optimise the number of peaks (also referred to as 'banks') in their wave-like schedule structure, as well as the detailed shape and structure of these banks. Initially, in the early 1980s, there was a trend at major airports towards many peaks, as a large number of banks in theory boosts both connectivity (as long as passengers care about waiting times during transfer) *and* productivity. Some airlines established up to eight inbound and eight outbound banks, which ensures a good match with the average roundtrip times of continental flights in Europe and the United States.

The pendulum swung back around the beginning of the new millennium, though, since airline managers found out that going for the maximum number of banks comes at the price of massive downsides.[10]

- Peaks in the hub schedule entail peaks in the utilisation of airport resources (e.g., ground handling staff, terminal space, baggage transfer systems, etc.) as well, resulting in capacity buffers for peak demand, and thus very poor productivity of these resources.

- Likewise, a wave-like schedule overloads the airspace during peak times, causing holding patterns for inbound aircraft, and thus delays and a deterioration in aircraft productivity.
- A high number of banks fosters short connecting times between inbound and outbound flights, close to the minimum connecting time, which is an operational performance parameter and characteristic of each airport or terminal. This is favourable, on the one hand, since it reduces idle time for both aircraft and passengers. On the other hand, short connecting times raise the risk of missed connections if inbound flights are delayed, plus cause inconvenience for passengers who can barely make the connection in a very large airport with long walking distances to cover. Both effects are highly counterproductive for hub airlines.
- Banks designed for short-haul connections have different efficiency requirements from connections for long-haul traffic. Combining them in a very dense hub pattern may result in compromises, thus hampering the efficiency of both the connection types.
- Hub carriers often try to assume a strategic perspective in designing their bank structure, for example by preparing the banks for expected future growth. This is risky, since growth may not occur as planned. The hub pattern may then be oversized, spreading the existing demand too thinly across the banks. If banks are not filled with a minimum number of flights and passengers, they become under-critical – i.e. every incremental bank dilutes connectivity instead of boosting it. In this case, getting the hub the right size and compressing the banks constitute the only way to regain a commercially and operationally sustainable structure (see Franke for comparison).[11]

Because of these downsides and risks, most airlines have in the meantime abandoned the ideal hub paradigm, a trend called 'de-hubbing' or 'de-peaking'. Instead, they have established hybrid structures designed to attenuate the above-mentioned downsides of peaky hub structures. At very large hubs (with more than 40 million passengers per annum), hub carriers have condensed their banks into some four mega-banks (e.g., Frankfurt), or completely switched to a continuous connecting mode ('random hubbing', e.g., British Airways at London Heathrow or Turkish Airlines at the new Istanbul International Airport).[12]

6.5.2 Options for mid-sized hubs: the meta-model

Opposed to very large hubs, home carriers in mid-sized hubs (some 10 to 40 million passengers per annum) do not own enough traffic volume to sustain continuous hubbing, so they still have to find the right bank structure to ensure profitable and stable operations. Besides fundamental drivers such as overall demand and transfer share, the ideal number and shape of banks depends on a variety of local, specific drivers, such as opening hours, geographical position relative to key markets or minimum connecting time.

Franke has demonstrated that mid-sized airports have a theoretical sweet spot in terms of bank numbers between five and seven banks (inbound and outbound).[13] The ideal number is determined, in the first step, by overall traffic volume and the transfer share of the home carrier at that airport (see Figure 6.2). This heuristic correlation (called the 'meta-model') is based on a simplified driver model, as well as on a benchmark across various European airports. In the second step, this initial recommendation for the ideal number of banks can then be adjusted to the specific drivers of the respective airport, such as its geopolitical position, the directionality of traffic flows or ratio between short-haul and long-haul flights.

The presumed mechanism behind the meta-model illustrated in Figure 6.1 is the following: the total passenger volume of a hub carrier (horizontal axis) determines the scale level of the hub. The more flights and passengers, the more banks can be filled. As explained earlier, the upper boundary for this correlation is a passenger number of around 40 million; hub carriers with more

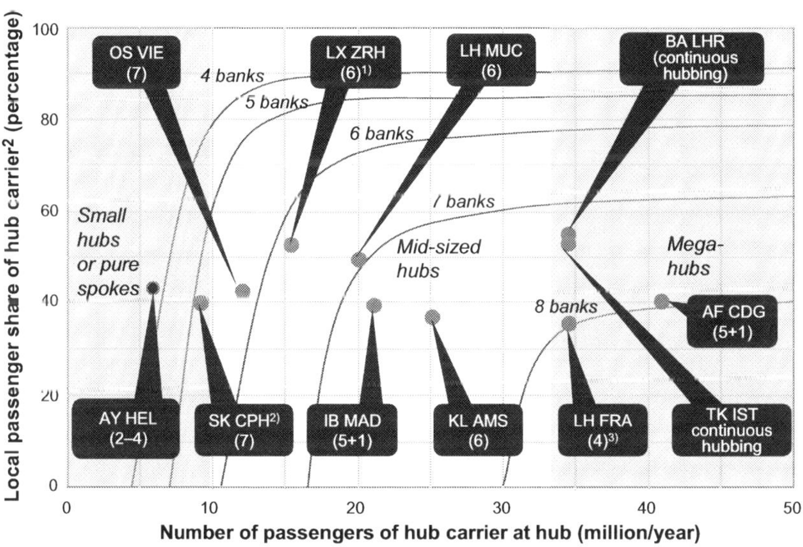

Figure 6.2 Heuristic calculation ('meta-model') of recommendable number of banks for mid-sized airports[a]

Notes: OS = Austrian Airlines; VIE = Vienna; AY = Finnair; HEL = Helsinki; LX = Swiss; ZRH = Zurich; SK = SAS Scandinavian Airlines; CPH = Copenhagen; LH = Lufthansa; MUC = Munich; FRA = Frankfurt; IB = Iberia, MAD = Madrid; KL = KLM; AMS = Amsterdam; BA = British Airways; LHR = London Heathrow; AF =Air France; CDG = Paris Charles de Gaulle; TK = Turkish Airlines; IST = Istanbul.

[a] Franke, 'Network design strategies'.
[1] Number of actually operated banks.
[2] Local pax share estimated.
[3] Overlapping mega-banks.

traffic are able to switch to mega-banks, or even continuous hubbing. The lower boundary for the heuristic lies around 10 million passengers; smaller carriers below this threshold cannot fully leverage the benefits of a wave-like bank structure, since they do not own enough volume to fill the banks.

The vertical axis shows the transfer passenger share of the hub carrier at this airport. With a transfer share of 0 per cent, the airport is by definition no hub, so no banks are needed; the schedule can fully focus on the timing preferences of the local (outbound) passengers. For a transfer share of 100 per cent, as the other – rather unrealistic – extreme, the preferences of local clients are irrelevant, since they do not exist. The bank structure needs to take only the needs of transfer traffic into account, which will result in a high number of banks in an ideal setting. In real life, typical transfer shares lie between 25 and 60 per cent, depending on the size of the local market; at hubs with a high-volume catchment, transfer shares are usually smaller than at hubs with a weaker catchment.

The non-shaded area in the centre of the matrix represents the segment for which the meta-model works best: the mid-sized hubs with a hub carrier owning between 10 and 40 million passengers per annum. The lines are curves with the same recommended number of banks; in general, the number grows from top to bottom and from left to right in the meta-model matrix. The shape of the curves imply that, if we start from a certain point on the horizontal axis (i.e., with a local passenger share of 0 per cent), the number of banks is determined by the volume only (e.g., slightly fewer than six banks for 10 million passengers, or slightly more than seven banks for 20 million passengers). If we now move up in the matrix on a vertical line (i.e., the overall volume stays constant, but the local passenger number goes up), the recommended number of banks decreases – at first very slowly, but after a certain point with a higher rate of decrease. This effect coincides with the 'knee'-like shape of the curves with constant bank number.

For example, a growing local passenger share for a constant volume of 20 million passengers initially pushes the recommended number of banks only very slowly from slightly above seven towards seven – a point that is reached at a local traffic share of 48 per cent. Since this point is at the same time the apex of the 'knee curve' for seven banks, a further increase of the local passenger share pushes the recommended number of banks with a growing change rate down to six, five and, ultimately, four (for a local passenger share of 90 per cent). The reason for this phenomenon is that, with local passenger shares of 50 per cent and above, the transfer traffic becomes increasingly irrelevant, and the hub carrier cannot fill more than a few banks.

Another interpretation of the 'knee curves' is that the same number of banks may be appropriate for quite different hub settings. For example, six banks are recommendable for a quite large airport with high local share (30 million: 77 per cent), but also for a much smaller hub with smaller local share (14 million: 20 per cent). At the latter airport, the home carrier is obviously able to compensate for its smaller volume with a higher share of transfer passengers, which is an intuitive trade-off.

As stated above, the bank number recommended by the meta-model is only a first step towards an efficient solution, as a comparison with the actual number of banks of mid-sized European airports underlines (see Figure 6.1). Some hub carriers match the recommendation of the meta-model (e.g., Swiss in Zurich, with six banks), whereas some operate significantly below (e.g., Iberia in Madrid, with five to six banks instead of above seven) or above the recommended number (e.g., Austrian Airlines in Vienna, with seven instead of close to six banks). The difference could indicate that the model is smarter than the respective hub carriers; the more likely explanation, though, it that Iberia and Austrian Airlines need to observe additional factors that are not reflected in the basic meta-model.

Therefore, the meta-model includes a second step designed to increase or lower the initial recommendation based on hub-specific drivers, such as geo-political position, the directionality of traffic flows and the ratio between short-haul and long-haul flights. For instance, Austrian Airlines has a rather low ratio between long-haul and short-haul flights, combined with a rather high share of mid-haul flights to Eastern Europe and the Near and Middle East. Both traffic characteristics drive the number of required banks up. In contrast, Iberia has a rim position in Madrid compared to the rest of Europe, with a relatively high share of long-haul business; both aspects drive the number of required banks down.

In addition to these hub specifics, it must be kept in mind that, at some hub airports, the number of installed banks is not absolutely clear, since the respective hub carrier operates one or more smaller peaks around noon or in the afternoon, rather difficult to identify as a bank.

Once the ideal number of banks has been determined and adjusted to the specific environment of a hub, the next level of optimisation is the fine-tuning of the shape and relative position of the banks (see Figure 6.3). For instance, in an ideal setting, a certain outbound bank should start no earlier than when the last flight from the preceding inbound bank has landed, because an overlap between both banks dilutes the theoretically attainable number of hits. In practice, it may nevertheless be appropriate to establish some overlap, thereby boosting the quality of connections (such as average connecting time), as well as the productivity of fleet and ground handling resources as a consequence of the smoothing of the peak effect.[14]

6.5.3 Top-down versus bottom-up approaches to hub planning

Another finding from the above-mentioned study[15] is that the strategic value of designing hub structures can be improved if a new perspective is added. The traditional way of planning network and hub structures is a one-way road: strategic and financial goals of an airline are translated into fleet and network

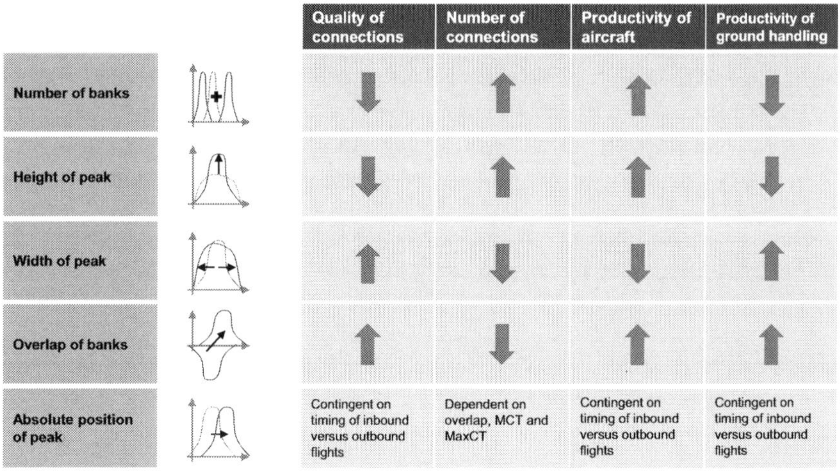

Figure 6.3 Impact of shape and relative position of banks on the key performance indicators of a hub[a]

[a] Franke, 'Network design strategies'.

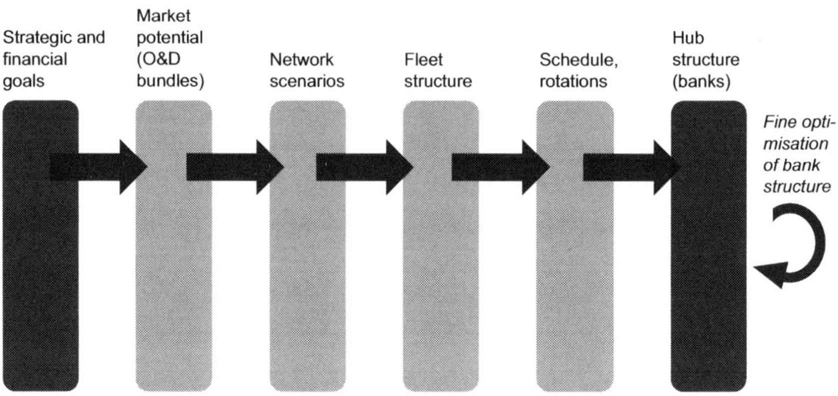

Figure 6.4 Outside-out/top-down approach to hub planning

scenarios, which will then be translated into a market schedule and rotational patterns. In this paradigm, the bank structure of the hub is perceived as the ultimate result of the fleet and network concept, standing at the very end of the planning process. This approach is illustrated in Figure 6.4.

This approach may lead to optimal results at the local level only, however, if not to inefficient results, since the hub structure can easily exert an impact on the overall profitability of the airline that is as large as the impact of fleet or network scenarios – or even larger.

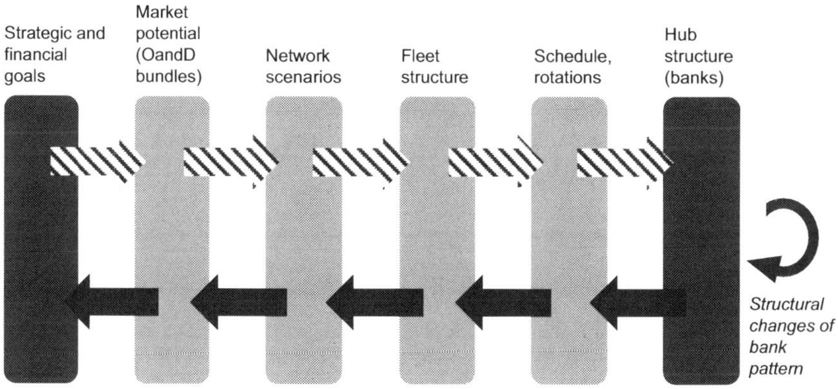

Figure 6.5 Inside-out/bottom-up approach to hub planning

Thus, a bank structure perceived as the natural outcome of a fleet and network concept bears the immanent risk of underestimating the complex interplay between strategy, demand, the fleet, the network and the hub structure. Sticking exclusively to a top-down planning approach may also end up in a conceptual dead end, since any changes of the hub environment cannot be played back into the planning cycle. For example, if the opening hours of a hub are reduced, it may be the best reaction not only to compress banks and stick to the previous network concept but also to calculate a new scenario with adjustments to network and market coverage. Adding that reverse, bottom-up perspective could establish the bank structure of a hub not just as *a result of* but also as *an input for* fleet and network scenarios.[16]

This insight leads to an advanced network and hub planning concept, featuring a reverse planning sequence: new hub scenarios are designed to accommodate major external or internal challenges, after which the effect on rotational patterns, fleet structures, network scenarios, market coverage, and – ultimately – the feasibility of strategic or commercial goals is simulated. This reverse sequence is shown in Figure 6.5.

6.6 Network and regulatory aspects

Although deregulation of aviation has been ongoing for more than four decades, air traffic is still a massively regulated business. Consequently, network design and operation are driven not only by economic factors and physical resources but by legal or regulatory restrictions as well. There are at least five major regulatory drivers:

- traffic rights;
- airports slots;

- opening hours of airports, night curfews;
- operating hours of crews; and
- flight standards, safety and security requirements.

Traffic rights: the transportation of passengers and goods has been subject to extensive regulation throughout the history of mankind, since transport has always been a source of wealth and power. Today air traffic rights are mostly negotiated on a bilateral level between countries, building on an entrenched system of *freedom rights* established for deep sea and inland water transport many centuries ago. Freedom rights start with level one (overflight rights for a foreign country) and culminate with level eight (some authors even define a ninth freedom, called *cabotage*, being the right to freely transport passengers and/or cargo *within* a foreign country). The following list provides an overview of all nine freedom levels (all statements for passengers are equally valid for cargo).

- **First freedom (overflight)**: the right of a carrier from country A to fly over country B.
- **Second freedom (stopover)**: the right of a carrier from country A to land in country B for technical reasons (refuelling, maintenance), but not for embarking or disembarking passengers.
- **Third freedom (outbound transport)**: the right of a carrier from country A to fly from country A to country B, embarking passengers at the origin and disembarking passengers at the (foreign) destination.
- **Fourth freedom (inbound transport)**: the right of a carrier from country A to fly from country B to country A, embarking passengers at the (foreign) origin and disembarking passengers at the destination.
- **Fifth freedom (outbound beyond or triangle transport)**: the right of a carrier from country A to fly from country A via country B to country C, embarking passengers or cargo at the origin and disembarking passengers or cargo at the destination (and vice versa: from country C via country B back to country A).
- **Sixth freedom (transfer transport)**: the right of a carrier from country A to fly from country B via country A to country C, with the option to connect passengers from country B to country C in country A, disembark passengers from country B in country A or embark passengers to country C in country A.
- **Seventh freedom (foreign transport)**: the right of a carrier from country A to directly fly between country B and country C, embarking and disembarking passengers at origin and destination [similar to sixth freedom, but without intermediate stop in home country A].
- **Eighth freedom (mixed foreign and inbound transport)**: the right of a carrier from country A to fly from country B with a stop in country B to country A.
- **Ninth freedom (foreign transport, or cabotage)**: the right of a carrier from country A to fly within country B.

Some political federations, such as the European Union, have designed multilateral agreements for their members, treating every member like a home carrier in that federation. Ironically, this cabotage right is not used very often, since operating in a foreign country against the national incumbent is a costly and expensive endeavour. Similar agreements may be closed in the future between other countries or groups of countries (e.g., between the United States and the European Union), but the respective negotiations are complex and have been unsuccessful so far.

Airport slots: due to limited airspace and airport capacities, airlines need to apply for so-called slots – i.e., time windows for take-off, flight and landing. Flight slots are assigned on a case-by-case basis by the respective air navigation service provider, while airport slots are granted on application by the International Air Transport Association. Officially, slots are granted biannually at IATA slot conferences and cannot be traded among airlines. Informally, there is a grey zone, since slots at very popular and congested airports such as London Heathrow (LHR) are tremendously valuable. In the past airlines have decided to acquire other airlines mainly to get hold of their attractive slot portfolio, a prominent example being British Midland Airways (bmi), which was purchased first by Lufthansa Group and then by IAG because of its sought-after slots in London–Heathrow (see Deprosse and Händel for details regarding slot management).[17]

Opening hours of airports (and night curfew): in densely populated regions, such as the United States, Europe and Southeast Asia, the noise emission of incoming and outgoing aircraft is perceived as a major nuisance for people living in the vicinity of airports. Thus, many municipalities or regulators restrict aircraft movements throughout the night to protect the health of their citizens. Usually departures and landings are not allowed during the night (typically 23:00 to 06:00), or at least are limited. In many countries this curfew regulation is combined with airport fee concepts penalising undesired movements – e.g., during the so-called 'shoulder times' operating hours, immediately before the beginning or after the end of the night curfew. Take-offs or landings during shoulder times are basically permitted, but the respective fees are much higher than during daytime.

Operating hours of crews: although aircraft can be operated as long as it takes to reach a destination within their flight range (provided that their maintenance requirements are fulfilled), crew members are much more restricted time-wise, in order to protect their health, and avoid fatal mistakes arising from fatigue.[18] Duty times and flight times alike are limited per day, per month and per year. Furthermore, many airlines have labour agreements with their crews that contain even more rigid limitations than the legal ones. Overall, the restrictions are so diverse on a granular level that they represent a significant constraint for network scheduling, short-term adjustments and disruption recovery.

Flights standards, safety and security requirements: there are a variety of legal standards defined to ensure utmost safety and security standards. The standards are developed by national aviation authorities, international industry associations such as IATA or the International Civil Aviation Organization (ICAO) or supranational entities, such as EASA. Examples for binding standards on a

European level are EU regulation 965/2012 for flight standards and AOC management, or EASA CAMO Part M for maintenance compliance.[19] Airlines are obliged by law to nominate postholders in charge of operations, who can be held liable if operational standards or safety requirements are not met.

Notes

1 Franke, 'Network design strategies'.
2 Franke, 'Network design strategies'.
3 Franke, 'Network design strategies'.
4 Franke, 'Network design strategies'.
5 Franke, 'Network design strategies'.
6 Franke, 'Network design strategies', emphasis in original.
7 Franke, 'Network design strategies'.
8 Franke, 'Network design strategies'.
9 Franke, 'Network design strategies'.
10 Franke, 'Network design strategies'.
11 Franke, M., 'Airline hub optimization: screening bank structures to boost hub performance', speech at German Aviation Research Society workshop prior to third European Aviation Conference, Amsterdam, 5 November 2014.
12 Franke, 'Network design strategies'.
13 Franke, 'Airline hub optimization'.
14 Franke, 'Network design strategies'.
15 Franke, 'Airline hub optimization'.
16 Franke, 'Network design strategies'.
17 Deprosse and Händel, 'Network planning and slot management'.
18 Franke, 'Network design strategies'.
19 Franke, 'Network design strategies'.

7 Impact of network design on airline performance

Ultimately, airlines aim at creating sustainable profit with the air transportation of passengers and/or cargo. Some airlines, especially state-owned flag carriers, may have very strong secondary targets, such as job creation, connecting their home country to the world or mere prestige; none of these targets can be achieved without a certain level of profitability, though. On a high level, the drivers of a profitable network have been discussed in the previous chapters:

- coverage of as many revenue streams (in markets/on O&Ds) as possible in a sustainable way;
- good fare mix between standard tariffs (business and first class) and special tariffs (economy), high average yield (RASK);
- deployment of most efficient aircraft type for a certain route (in general, larger aircraft provide a better unit cost position – provided that their larger capacity can be filled);
- sustainable position on the so called 'S curve' in key markets – either in a sustainable niche with limited capacity, but very good efficiency, or as dominant carrier with an attractive frequency pattern and large aircraft;
- achievement of good seat load factor (with 'good' being – as an experience-based rule of thumb – 70% < SLF < 95% for a long-haul flight and 75% < SLF < 95% for a short- or mid-haul flight in a hub and spoke setting);
- reasonable utilisation of assets, especially of aircraft (with 'reasonable being > 10 block hours per day for hub and spoke carriers and > 12 block hours per day for low-cost carriers, both on continental routes, and > 14 block hours per day for long-haul operations); and
- lean operations, especially in the hub airport (for hub and spoke carriers).

Additional marketing levers and revenue streams (e.g., ancillary services, upgrades, service fees for other airlines) are suited to generating more profits, but they are not core elements of network management.

On a more granular level, there are additional correlations between network design and commercial success, on the one hand, and operational stability, on the other.

7.1 Impact of network design on commercial success

For low-cost carriers, the correlation between network design and commercial success is rather simple, since the inherent interdependence of resources – and thus the complexity – is very limited. Therefore, the levers described in the introduction of this chapter are mostly sufficient to steer the business and ensure profitable operations. If a route, or even a base, does not operate profitably for more than a few weeks, a low-cost carrier may decide to skip that route or close the base.

For hub and spoke carriers, however, this correlation is more complex. A basically efficient network structure may still generate losses. The main reason for this – at first glance counterintuitive – effect is the complexity of the hub. A hub is a core element of the production model of a hub and spoke carrier, determining considerable parts of the carrier's operating costs. On the other hand, the hub structure exerts a considerable impact on how travellers perceive the service offering of an airline, and which travellers the airline can serve at all. In this respect, both network and hub determine the revenue and cost position of an airline at the same time.

As already described in Chapter 6, section 6.5, a chosen hub structure entails a variety of compromises and trade-offs – even more than in the case of the rest of the network. For instance, a very peaky bank structure may ensure good connectivity, and thus a high revenue potential for the entire network. On the other hand, high peaks in the daily flight pattern could lead to congestion, worsening the operational stability and the productivity of resources in the air and on the ground.

Besides understanding connectivity as a mere performance indicator of a hub, it is crucial to know *which kind of markets* or O&Ds are connected in that hub. Most major airlines operate major hubs – i.e., they go for scale. Although scale is, basically, good for a sustainable cost position, it entails a certain risk: the respective hub and spoke carrier ideally needs to cover each and any market or O&D within reach of the hub, which is usually referred to as '360 degree approach'. In its most literal interpretation, '360 degrees' implies not only 'many' (markets/O&Ds) but also 'evenly spread around the hub'. Both prerequisites can be very difficult to achieve, if not counterproductive.

7.1.1 Downsides for under-critical hubs

A negative backlash type of effect may be most likely to occur when a hub and spoke carrier operates a hub that is still under-critical in terms of traffic volume. To achieve sufficient growth with a 360 degree model, the respective hub carrier has almost no other option than to walk down the following two paths:

- operate smaller aircraft (than the home carrier of a larger hub could do) on short-haul and long-haul routes alike to ensure sufficient seat load factors despite a growing number of frequencies; or

- build a dense network of feeder/de-feeder flights with narrow-body aircraft on a domestic and continental level.

Both approaches entail massive risks. Deploying on average smaller aircraft in a hub than comparable competitors establishes a structural unit cost disadvantage – a permanent downside both on point-to-point and connected services. If there are two hub and spoke carriers with adjacent hubs (e.g., Air France in Paris Charles de Gaulle, Lufthansa in Frankfurt Rhein-Main and British Airways in London Heathrow), they compete for the same transfer passengers on many O&Ds between the Western and Eastern Hemispheres. The hub carrier with the largest fleet and generally largest aircraft is able to offer cheaper flights and attract disproportionately more demand than its competitors.

Of course, a hub carrier can grow its business with the 360 degree approach, provided that its pockets are deep enough to bridge the profitability gap described above. Growing volume fills larger aircraft, reducing the scale disadvantage step by step. The investment can be sizeable, however, not to forget the cost applicable to larger aircraft (if not available at other spots in that network).

7.1.2 Build-up of continental feeder/de-feeder network

Building a dense domestic and continental feeder network is indispensable for the efficient operation of attractive long-haul services – especially if the local demand reservoir (catchment) of a certain hub is limited. The success of a hub is built on an economically robust blend between the local generation and cumulation of demand. Many short- and mid-haul services of hub and spoke carriers make sense only from a network standpoint; on a stand-alone basis, they are unprofitable. Thus, it is crucial to establish a sustainable balance between (hardly profitable, or even unprofitable) feeder flights, slightly profitable short-haul flights and (often highly profitable) long-haul flights. This balance is very difficult to achieve during the growth phase of an under-critical hub.

7.1.3 Types of connections realised in a hub

Figure 7.1 illustrates the four possible combinations of traffic types (inbound and outbound) in a hub. Reflecting the required density of the feeder network, featuring short- or mid-haul flights, the transfer from domestic or continental to domestic or continental flights is usually by far the dominant scenario in a hub (bottom left box). Unfortunately, this combination is the least profitable – if not unprofitable scenario. Carriers operating connected services (via their hub) on a short- or mid-haul O&D are usually forced to offer discounted fares, since passengers are nowadays able to find non-stop connections on the vast majority of continental routes, accepting connected services only at reduced fares.

Thus, a large number of short-haul flights are needed to create a base load for the hub and ensure the feed or de-feed for the profitable long-haul flights.

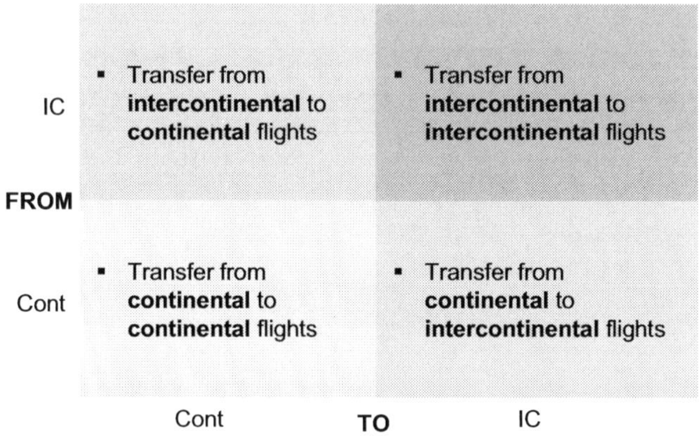

Figure 7.1 Combinations of different traffic types in a hub[a]

[a] Franke, M., 'Airline case study from FATC project database', 2014 [not published].

This large number translates into an even higher share of continental-to-continental connections, since connections are obviously correlated in a quadratic way with the number of underlying flights. In essence, the bank system of a hub benefits from the sheer mass of short-term flights – since well-filled banks can be operated more efficiently in terms of connectivity and productivity than banks with only a few flights – but, at the same time, the bank structure is congested and distorted by the least profitable connections.

The top left and bottom right boxes in Figure 7.1 stand for the second most frequent case in a typical international hub: the connection between short-haul and long-haul flights, with the continental or domestic flight being the feeder (or de-feeder) service for the intercontinental flight (e.g., Philadelphia via New York to Frankfurt and vice versa, or Milan via London to Tokyo and vice versa). This combination is less frequent than the case above, but considerably more profitable. First, the long-haul flight has a higher yield level by itself, and, second, the availability of non-stop flights is much lower on intercontinental O&Ds than on a continental level, so that carriers are not forced to offer structural discounts for their connected services.

The most attractive case for a hub and spoke airline – in commercial terms – is the top right box in Figure 7.1: the connection between intercontinental and intercontinental flights (sixth freedom hub traffic) – e.g., Chicago via Amsterdam to Beijing, or New York via Dubai to Melbourne. Unfortunately, this case is the least frequent one, with a share of some 5 to 15 per cent, for instance, in continental Europe.

7.1.4 Case study for optimisation of bank structure

Figure 7.2 shows the results of a real-life network and hub optimisation case. In a major European hub airport (with over 40 million passengers per year),

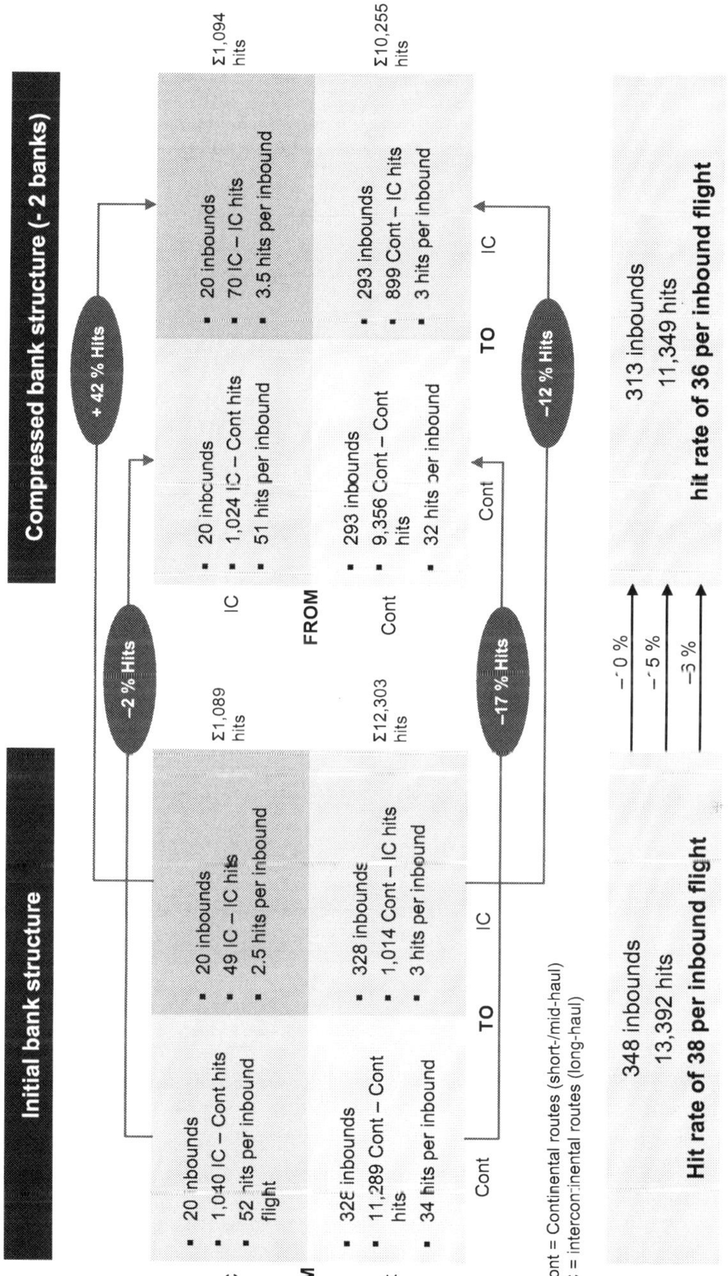

Figure 7.2 Case study on improved network efficiency through bank compression in a major European hub[a]

[a] Franke, 'Airline case study from FATC project database'.

the home carrier realised that the aspired growth trajectory was not sustainable, since profitability turned out to deteriorate more and more. The reasons were the ones discussed earlier in this chapter (subsection 7.1.1): the number of long-haul flights was too small compared to the number of short- and mid-haul flights, ending up with a bank structure favourable for the (many) continental-to-continental connections, but unfavourable for the much more profitable continental-to-intercontinental and intercontinental-to-intercontinental transfers. The connections between continental (feeder) flights blocked the system, preventing the bank structure from being optimised for the more profitable intercontinental connections.

The initial number of banks operated by the home carrier in that hub was eight (both inbound and outbound), which was – in theory – a pattern fostering both connectivity and productivity of assets. With around 16 operating hours per day and an average roundtrip time of two hours, eight banks ensured an ideal rotational pattern with good utilisation of aircraft and crews. Furthermore, a high number of banks leads to high connectivity in an ideal setting. Unfortunately, the setting for this carrier was not ideal: compared to mega-hubs, the number of flights per bank was much smaller, spreading the traffic volume too thinly across the entire pattern. In practice, having a high number of banks with too few flights per bank turns out to be counterproductive, since they decrease the overall connectivity of a hub instead of increasing it.

7.1.5 Correlation between number of banks and asset utilisation

Likewise, having a large number of banks compromises rather than fosters high productivity, since the connections get very tight. The time window for the aircraft being prepared for the next flight (turnaround), as well as for the passengers moving to their onward flight, may become so small that connections do not work any more. In this case, the operational instability induced in the system tends to absorb the theoretical advantage of many nicely synchronised aircraft rotations.

Consequently, most major carriers operating mega-hubs have left the eight-bank system behind and consolidated the eight banks into four mega-banks (e.g., Lufthansa in Frankfurt). Given their superior volume and scale, the loss of connectivity is tiny. On the other hand, four mega-banks give these carriers much more leeway to design their rotational patterns efficiently, without too many tight rotations and connections that induce operational instability. Some other major carriers, operating hubs with even larger catchments, have gone beyond this solution and perform 'continuous hubbing' – a pattern in which the volume is so high that single peaks are very difficult to distinguish.

In the concrete case mentioned above, the carrier operating the mid-sized hub decided to walk down that path as well, compressing the bank pattern from eight to six and cutting away at the same time some loss-making flights. Furthermore, as stated above, the carrier rearranged its bank structure in that hub in a way that it was ideal for the most profitable connections (intercontinental to intercontinental, continental to intercontinental, intercontinental to continental), but not for the least profitable transfers (continental to continental). This comprehensive reset of the hub structure led to increased revenues and profits despite a slight drop in flights, connections and passengers.

Figure 7.2 shows the changes achieved through this move: initially (i.e., in the left matrix), 328 continental and 20 intercontinental inbound flights per day produced 11,289 viable connections ('hits') between continental flights (34 hits per inbound flight), 49 connections between intercontinental flights (2.5 hits per inbound flight), and 2,054 mixed connections (intercontinental to continental: 52 hits per inbound flight or vice versa: 3 hits per inbound flight).

After the reshuffling and compression exercise (i.e., in the right matrix), the number of long-haul inbound flights remained unchanged, while the number of short- and mid-haul flights was slightly reduced (by some 11 per cent). Through this reduction, together with the shifted focus (to bank structures designed instead for high-value connections), the number of short-haul connections slumped by 17 per cent – significant, but uncritical because of the poor profitability of the served O&Ds. The less dense feeder/de-feeder network also had a negative impact on the hit rate for mixed connections: –2 per cent for intercontinental to continental transfers, –12 per cent for continental to intercontinental transfers; nevertheless, even this loss of volume was made up for by the improved quality – and thus revenue – of the connections. A nice upside was finally achieved on the highly profitable intercontinental connections, with 42 per cent more hits.

7.1.6 Results of the hub reshuffling initiative

All in all, the carrier was able to improve the profitability of its hub (and, thus, its network) despite the significant consolidation and reduction of traffic volume. The overall key indicators changed as follows:

- number of inbound flights: –10 per cent
- number of hits: –15 per cent
- number of hits per inbound flight: –6 per cent
- number of deployed aircraft: –7 per cent
- number of O&D passengers going through the hub: –7 per cent
- number of destinations served by that hub: –2 per cent (loss of continental destinations only)
- O&D revenue: 0 per cent
- number of local passengers: +1 per cent
- local revenue. +4 per cent
- overall profitability (profit contribution 3 margin): +6 per cent

The figures show another favourable side effect of the hub realignment: the bank structure has been optimised not just for the valuable intercontinental connections but also for the local passengers (from the catchment of the hub) who are also high in value, since they pay undiscounted fares for attractive non-stop flights. Ultimately, the hub carrier generated 6 per cent more profit and 4 per cent more (local) revenue with 7 per cent fewer aircraft and 10 per cent fewer inbound flights – a very successful example of profitable slimming.

7.1.7 *Impact of hub structure on the profitability of destinations*

Another impact of network and hub structure design on profitability is reflected in the margin of a hub and spoke carrier for each destination served. Having the S-curve effect in mind (see Chapter 6, section 6.3), the intuitive conclusion may occur that high-volume markets basically generate higher margins than thin routes, since they support much higher frequencies. This is not the case, however, since 'high' frequencies are a necessary but by no means sufficient criterion for economic success on a certain O&D or route.

Instead, it is the frequency position *relative to other airlines* on this route that drives the profitability of services, plus the overall intensity of competition. For example, ten daily flights by carrier A on a continental trunk route can be 'low' if carrier B offers 12 and there is a cut-throat competition on this route; carrier A may end up with a poor economic result. On the other hand, three daily flights on a secondary continental route may lead to a very satisfactory margin for carrier A if carrier B is a niche player with only one daily flight, and is not at all in a position to challenge carrier A.

Figure 7.3 shows the real-life case example of a mid-sized European hub carrier. Profitability (in terms of contribution margin 2; see network profit calculation scheme in section 3.3) as well as number of frequencies are plotted per destination, sorted in the order of traffic volume (measured in onboard passengers of the respective carrier: large routes on the left, small routes on the right). It becomes evident that there is indeed a strong correlation between volume on a route and number of frequencies (the grey dashed line, mainly growing from right to left), but there is *no correlation at all* between market size and profitability! The former effect is a clear result of the carrier observing the S-curve effect, while the latter phenomenon is attributable to the impact of competitive intensity.

7.1.8 *Clustering of destinations by O&D size and profitability*

To account for competitive intensity, the plot can be divided into four clusters:

- *large (continental) trunk routes* (in this concrete case: traffic volume of the carrier > 200,000 passengers p.y., > 20 flights p.w.)
- *medium large routes* (> 100,000 passengers p.y., > 15 flights p.w.)
- *medium small routes* (> 50,000 passengers p.y., > 10 flights p.w.)
- *small routes* (< 50,000 passengers p.y., < 10 flights p.w.)

The thresholds between the clusters do not spring from a universal mathematical formula but have been selected in a way that made the different competitive

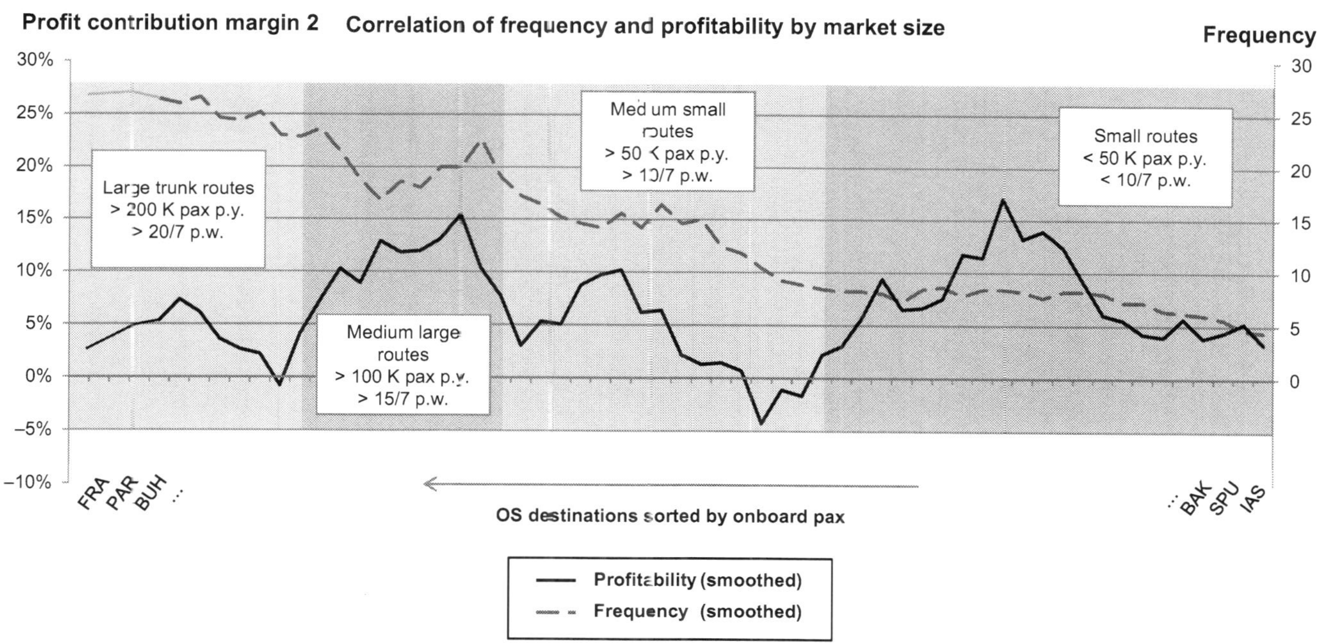

Figure 7.3 Interplay between market size, frequency and margin for a European hub carrier[a]

[a] Franke, 'Airline case study from FATC project database'.

environments most obvious. The thresholds could be chosen differently, and will in any event be different for other carriers.

The resulting clusters can nonetheless be interpreted in a quite intuitive way.

- On *large trunk routes*, competition is immense, since every carrier operating in that geography is forced to serve the respective destinations. Consequently, it is almost impossible to differentiate by frequency or service level as a single carrier; instead, the only parameter that makes a difference is price. Thus, relative profitability (i.e., margin) on trunk routes is usually mediocre, while the absolute profitability is high thanks to large volumes.
- *Medium large routes* are highly profitable across the board for this carrier, because they are small medium (or even small) routes for other carriers. The reason for this differentiation may be a historic stronghold or any other kind of special competence in this market. Consequently, the carrier dominates the market with a superior number of frequencies, unleashing the full potential of the S-curve effect. Absolute profitability may be smaller than in the first case, though, since the markets are smaller.
- *Medium small routes* are the flipside of the previous case, since they are large medium routes for one or more competitors. Consequently, the carrier under discussion is only a niche carrier, dominated by one or more incumbents. It will suffer from the negative S-curve effect, and realise only mediocre margins (if not losses). The absolute profits are also poor, reflecting the relatively small market sizes.
- On *small routes*, margins are often excellent, since competition is either weak or non-existent; the active carrier is often a monopolist with a historically unique position in the respective market. Absolute profits are of course small, though, on account of the small size of the markets.

7.1.9 Lessons from the clustering for network design and scheduling

If the described carrier wishes to expand or consolidate its business, the plot in Figure 7.3 provides valuable evidence for such a strategic decision. The first cluster (*large trunk routes*) is the backbone and 'cash cow' of the network and should be operated in any event, since it provides base load for the (continental) network and reasonable profits. If the incremental profit is positive, additional capacity could be assigned to routes in this cluster.

Product features and service quality should be mostly in line with those of competitors on these routes, but the main success factors on trunk routes are capacity and attractive timing of flights – especially those matching the needs of business travellers. The latter are not overly price-sensitive, so price is usually not a major differentiator on trunk routes; most carriers pursue a pricing strategy that counts on matching the fares of the dominant carrier in a certain destination (e.g., American Airlines in Dallas/Fort Worth, or British Airways in London Heathrow).

The second cluster (*medium large routes*) contains most of the 'stars' of the destination portfolio of a certain carrier, since the carrier dominates these markets, and generates reasonable profits. If possible, the business on these routes should be expanded, resulting in more capacity as long as the saturation point on the S-curve has not yet been reached, and new routes should be developed into this cluster (from the current *medium small routes* cluster). Both strategies contain considerable risk, however, because medium large routes are highly attractive for *every* airline able to operate on these routes, so the competitive intensity may be high. Massive capacity additions or even price wars may lead to a fierce blood-shedding, spoiling the respective market at least for a certain period (cf. case study in Chapter 8).

The *medium small route* cluster contains 'poor dog' routes, those low in margins or even loss-making. To some extent, these routes may be needed to achieve the 360 degree coverage and a certain critical mass in the airline's feeder network; this segment should be carefully screened for candidates that could be taken off the network without any negative network effect, though. Structural loss makers should be eliminated if possible. On the other hand, there may be routes that can be developed into a sustainably profitable position, for instance by adding capacity and migrating from a niche into a dominant player position.

That is particularly true if that route is a *medium small market* for this carrier only, but a *medium large market* in general. The required investment for this move may be significant, though, and the current incumbents may fight back at any price, since this route is a highly profitable, medium large type of market for them, which they cannot afford to lose. In all other cases – overall market being too small for a 'star' positioning or expansion of the current position being physically not viable – closing 'poor dog' routes can be the most reasonable strategy. As demonstrated earlier in this chapter in the hub case study, smartly downsizing a hub or a network can definitely lead to more revenues and profits – provided that the slimming is performed in the right way and based on the right analyses.

The fourth and final cluster (*small routes*) features a variety of 'question mark' routes, with small volume but relatively high margins, due to monopolistic or quasi-monopolistic structures. Although these routes will not justify a determined growth strategy, it may still make sense to sustain them, and grab all the cash they are able to deliver. Even in small markets there may be a few clients for long-haul destinations; if one carrier has exclusive access to these markets, it at the same time has exclusive access to the relatively high long-haul fares these clients are ready to pay, connecting them via its hub.

The entry of a second player, especially of a low-cost carrier, could very quickly spoil that game, but many of these specialist markets with a historic incumbent are too small or too remote even for a low cost carrier. As long as no other airline interferes, the current incumbent should stay and extract the available revenue from the market.

7.2 Impact of network design on operational stability

The previous chapter has already pointed out that network patterns looking great in theory may not work out well in practice. This is true not only in terms of profitability but also with respect to operational viability or stability. Thus, every network or hub structure should be carefully checked for operational stability to ensure a sustainable business case. There are various scenarios in which network structures may compromise operational performance.

- *Too short turnaround times*: a quick turnaround is a key success factor for high aircraft and crew utilisation, since it enables more productive hours of the most expensive resources. For operations, however, a quick turnaround is extremely challenging: first, a turnaround process comprises a variety of internal and external services (such as fuelling, cleaning and catering), which are very difficult to manage under high time pressure. If the airline's ground handling manager puts too much pressure on the service providers, quality may suffer; if he puts too little pressure on them, the process may be too slow and create delay. Second, if the network is built on very ambitious turnaround times across the involved airports, every little delay may create massive knock-on effects, propagating the delay throughout the entire network, and causing an inherent instability.
- *Too tight connections (passenger perspective)*: similar to the previous point, short connections are in theory great to reduce waiting times between inbound and outbound flights, resulting in an improved aircraft and crew utilisation. Consequently, airports – which are at least partly responsible for connecting times – even promote the short connections achievable with their infrastructure layout and transfer processes. The KPI for this ability is called minimum connecting time and describes the average quickest transfer time for a traveller deboarding from his inbound flight A, crossing the terminal and boarding his outbound flight B. Quite obviously, the required transfer time depends, on the one hand, on the personal fitness and sense of orientation of a certain traveller; an elderly or disabled person will be much slower than a young and frequent traveller. On the other hand, the transfer time is driven by the distance between inbound and outbound gates, the overall layout and technical devices such as people movers or horizontal escalators. At mega-airports with several concourses and terminals, the minimum (as well as average) connecting time must by nature be longer than in a mid-sized airport with only one terminal. If an MCT is overly optimistic, many passengers and baggage items will get lost during their transfer (i.e., they will miss their connection or cause delay).
- *Too tight connections (aircraft perspective)*: if a flight schedule is too tight, every little irregularity may cause a substantial delay in the system. Issues such as bad weather conditions or congested airspace transform from an issue on one route into a bigger issue for a rotation or even a sub-network. On the other hand, operational buffers such as extended waiting times between

outbound and inbound flights are highly effective, but also vastly expensive. The sweet spot seems to be located in the middle between the two extreme positions: ambitious schedules with tight connections are absolutely needed to achieve high asset utilisation and thus competitive unit cost positions, but if the schedules are too tight the resulting operational instability will dilute the benefits of a productive schedule.

- *Flight departures or arrivals scheduled too close to shoulder times or night curfews of the respective airports*: making use of the entire operational time window of an airport with night flight ban improves the leeway for an efficient rotational pattern, potentially boosting aircraft and crew utilisation; it may very likely cause significant irregularity cost, though. If a delayed arrival leads the respective flight into the shoulder time of the destination airport, the landing fee may drastically rise, penalising the unwanted arrival during night-time. If the delay leads the flight into the night curfew of that airport, the resulting cost may even be higher: the flight must be diverted to another airport without a night ban, leaving the involved carrier with significant additional costs, such as hotel accommodation or bus transfer for stranded passengers, as well as dead-end cost for crews running out of their duty time – not to speak about the additional landing fees and transfer cost for the aircraft.

The above list is far from comprehensive; there are many more examples of critical trade-offs between efficient network or rotational patterns and operational stability. The learning is, in any event, that network designers and flight schedulers should proactively seek the advice and approval of operational managers to avoid scenarios in which theoretically smart network structures heavily backfire on them by inducing considerable irregularity cost. In essence, operational robustness and the reliability of transportation networks are values that should never be underestimated.

8 Case study on levers and effects of network management

This chapter concludes the first part of this book with a practical example. This example is not taken from real life, but it features a realistic problem with realistic data. The setting is a very basic network management problem:

- a single route (short-haul: 'Spoke Town – Network City') with a given O&D passenger flow
- three airlines (carriers A, B, C), owning two different types of narrow-body aircraft each...
- ...with different unit cost positions...
- ...but identical yield potential (for simplification)

The three competitors behave as in real life, trying to attract a sustainable market share with an attractive frequency pattern, assigning capacity to the expected demand in an efficient way and, ultimately, maximising their profit. They know the S-curve effect and act accordingly.

Table 8.1 provides the basic data and assumptions for the case study. Carrier A is the hub carrier and dominant player in Network City; carrier B is a traditional hub and spoke carrier, but only with an operational base in Spoke Town (no hub in Network City or Spoke Town); carrier C is a low-cost carrier with a home base in Spoke Town. Since the distance between Spoke Town and Network city is only 500 km (one way), all three carriers deploy narrow-body aircraft only (each of them owning two different types of single-aisle aircraft). The choice of the aircraft type determines the unit cost of the respective carrier on that route (i.e., CASK), based on the following simplified assumptions:

- aircraft with larger seat capacity feature lower unit costs than aircraft with less capacity;
- the listed seat capacities are only representative of the aircraft types; in real life, the seat capacity (as well as configuration) varies widely between carriers;
- the low-cost carrier C achieves lower unit costs than the two hub and spoke carriers A, B; and

Table 8.1 Basic facts for network management case study[a]

Case example: O&D 'Spoke Town' – 'Network City'				
Basic facts				
Distance (one-way, km)	500			
Flight time (one-way, hrs)	1			
Turnaround time (hrs)	0.75			
O&D demand (pax p.y. one-way)	365,000			
Pax p.d. (one-way)	1,00			
Max. operating hours p.d.	17			
Key data of carriers				
	Yield (US$/RPK)	**Aircraft**	**Seat capacity**	**Unit cost (US$/ASK)**
Carrier A (hub carrier in Network City)	0.08	A321	184	0.055
	0.08	A320	150	0.060
Carrier B (base in Spoke Town)	0.08	B737–500	160	0.060
	0.08	A319	120	0.065
Carrier C (base in Spoke Town)	0.08	B737–800	170	0.050
	0.08	B737–700	150	0.058

[a] Franke, 'Strategic network management'.

- all other cost differentials (e.g., overhead, ground handling, etc.) are not taken into account.

8.1 Basic assumptions for the case study

To simplify the calculation and focus on operational efficiency rather than on commercial drivers (marketing, product design, pricing, etc.), the achievable yield (revenue per RPK) is assumed as being identical for all three carriers. In real life, the yields would of course be different. All yield, unit cost figures and aircraft seat capacities are purely fictitious, though not unrealistic compared to real airlines. The demand on this O&D (1,000 passengers per day one-way) is notional.

If Spoke Town to Network City were a new route, the three carriers would have to build their operations from scratch and estimate their initial market share – based on their position in similar markets, as well as on their appetite for risk and investments. This new entry for all competitors would usually not lead to an equilibrium situation, and, if so, only by mere coincidence. More probably, the three carriers would over- or underrate their attainable market share, creating overall over- or undercapacities in that market. Whereas undercapacity does not cause too much harm for the carriers (only some potential clients stay on the ground, and prices probably go up), excess capacity is very likely to cause economic problems for one or even all carriers (see also below).

Since the start-up scenario is not very frequent any more in our highly mobile and globalised world, our example instead starts from an equilibrium point, the latter being defined by a match between expected and actual demand (see Table 8.2).

For our case study, we assume that carrier C (the low-cost carrier based in Spoke Town) has invested most aggressively into this market, achieving a 42 per cent market share (i.e., 420 passengers p.d. one-way). Carrier A, the hub carrier in Network City, seems to be less enthusiastic about this O&D (maybe the market is too weak, or the network contribution is poor), accepting a market share of only 25 per cent. Carrier B absorbs the remaining 33 per cent market share.

To uplift the respective 250 passengers, carrier A decides to schedule two daily roundtrips with its smaller (narrow-body) aircraft type. The resulting capacity of 300 seats (one way) translates into a seat load factor of 83 per cent, which is basically good enough to make money in a structurally profitable market, as in the case study. In this context, 'structurally profitable' means that the yield per RPK is higher than the unit cost per ASK of every single aircraft type available for this route; the only way to generate losses in this case is to generate poor load factors. The total costs, revenues, and profit of carrier A are calculated as follows:

Operating costs = unit cost (A320) per ASK x total seats (one-way) x flight distance (two-way) = US$0,060/ASK x 300 seats x 500 km x 2 = US$18,000

Ticket revenues = yield (on this O&D) per RPK x passengers (one-way) x flight distance (two-way) = US$0,080/RPK x 250 passengers x 500 km x 2 = US$20,000

Operating profit = revenues – costs = US$2,000

Carrier A does not have any other realistic option, since fewer roundtrips would lead to undercapacity (not all passengers willing to fly with carrier A could be uplifted), while larger aircraft or more roundtrips would generate overcapacity, penalised by poor seat load factors and lower profits.

Carrier B goes for three roundtrips with its smaller aircraft type, which is rewarded by a very good seat load factor of 93 per cent. This choice leads to costs of US$23,400, revenues of US$26,664 and a profit of US$3,264 (see Table 8.2). Different from carrier A, carrier B *does* have alternatives in this setting: it could, for instance, switch one of the three roundtrips with its smaller aircraft type (A319) to its larger aircraft (B737–500). This scenario would lead to the following key figures:

- Overall seat capacity = 400 seats (one-way)
- Overall seat load factor = 83%

Table 8.2 Base case: equilibrium point[a]

		Target market share	*Pax p.d.*	*Roundtrips p.d.*	*Seats p.d. (one-way)*	*Overall seat load factor*	*Total oper. cost (US$)*	*Total cost per airline (US$)*	*Total ticket revenue*	*Total revenue per airline*	*Oper. profit per A/C (US$)*	*Profit per airline (US$)*
Carrier A	A321	25%	250	0	0	83%	0	1,8000	0	20,000	0	2,000
	A320			2	300		18,000		20,000		20,000	
Carrier B	B737–500	33%	333	0	0	93%	0	23400	0	26664	0	3264
	A319			3	360		23,400		26,664		3,264	
Carrier C	B737–800	42%	416.7	1	170	89%	8,500	25,900	12,058	33,336	3,558	7,436
	B737–700			2	300		17,400		21,278		3,878	
	Total	100%	1,000	8	1,130	88%		67,300		80,000		12,700

[a] Franke, 'Strategic network management'.

- Total operating costs = US$25,200
- Total ticket revenues = US$26,664 (unchanged)
- Profit = US$1,464

This alternative scenario is obviously inferior to the initial one calculated in Table 8.2, as the overall profit is lower. The scenario could have been beneficial, since the drop in seat load factor could have been compensated for by the better unit cost position. This is obviously not the case, however, and a quick estimate shows why: the loss of overall seat load factor would be too massive (–10 percentage points) to be made up by the weighted unit cost advantage of only 3.1 per cent. Consequently, the operating costs go up (by 7.7 per cent), not down, and so does the profit.

Carrier C decides to schedule one roundtrip with its larger aircraft type and two with its smaller narrow-bodies, resulting in a seat capacity of 470 seats and a good seat load factor of 89 per cent. Operating costs sum to US$25,900, ticket revenues to US$33,336 and profit to US$7,436.

As an alternative, carrier C could have chosen to deploy its smaller aircraft type only (i.e., three roundtrips with B737–700), following the example of carrier B. The commercial results would be the following:

- Overall seat capacity = 450 seats (one-way)
- Overall seat load factor = 92.7 per cent
- Total operating costs = US$26,100
- Total ticket revenues = US$33,336 (unchanged)
- Profit = US$7,236

Opposed to the case of Carrier B, the reduction of capacity (i.e., deploying the smaller aircraft type only) would not add but eat up profit. Again, the effect can be easily explained: operating one roundtrip with the larger aircraft type lowered the weighted unit cost of carrier C by 4.5 per cent, at the expense of only 4 percentage points in seat load factor. It becomes very clear that, even in this tiny and very limited network management case study, there are various choices, and it is by no means trivial to identify the most favourable option.

8.2 Macroeconomic view on market efficiency

From a macroeconomic perspective, not only is the single profit per airline interesting but so is the total profit generated by all carriers in that market. In the case of the equilibrium scenario calculated in Table 8.2, the profit sums up to US$12,700. This sum can be interpreted as an indicator for the overall efficiency of the scenario. The question is now: is US$12,700 a good or a bad result?

A plausible reference case is a monopoly scenario. A monopolist replacing the three competitors on the respective O&D would be in a better position than each of the three carriers in Table 8.1, since he could benefit from scale effects in assigning capacity to demand (with growing volume, the cut-off risk generated by inflexible capacities per aircraft diminishes). Furthermore, the monopolist could choose the cheapest available aircraft type; in this case study, this would be the B737–800 from the fleet of carrier C. With these assumptions, the monopolist scenario delivers the following results:

- Flight schedule: six roundtrips with B737–800
- Overall seat capacity = 1,020 seats (one-way)
- Overall seat load factor = 98 per cent
- Total operating costs = US$51,000
- Total ticket revenues = US$80,000
- Profit = US$29,000

The resulting profit is significantly higher than that of the initial equilibrium scenario with three competitors, due to scale effects and the consequent use of the cheapest aircraft type. Furthermore, a monopolist could live with a very high seat load factor such as 98 per cent, whereas, in a competitive environment, seat load factors above 95 per cent are often unsustainable, since they tend to dilute overall revenues of the respective flight (cf. rules of thumb in Chapter 3, section 3.2).

Compared to this monopolistic reference case, the initial equilibrium case with competition achieves US$12,700/US$29,000 = 43.8% of the reference profit. All other competitive scenarios discussed below can be compared to that macroeconomic efficiency level.

8.3 Non-equilibrium scenarios

We will now discuss two *non-equilibrium scenarios*: a *capacity expansion* and a *price war*. In the first scenario, carrier A, the hub carrier in Network City, decides to be more ambitious on this O&D, and add capacity to attract more demand. Counting on the S-curve effect, the carrier adds one daily roundtrip with its larger aircraft type to make its frequency pattern more compelling and boost its market share. The underlying assumption is that the additional frequency makes the offering of carrier A more compelling for passengers, but only for passengers already present in that market. Thus, carrier A is able to entice clients away from carriers B and C, but not to win new clients beyond the current market potential (1,000 passengers p.d. one-way). For this case study, we assume that carrier A wins 5 per cent market share from both carrier B and carrier C. The key figures of this scenario are depicted in Table 8.3.

The figures in Table 8.3 demonstrate that the capacity addition increases the market share of carrier A, but does not pay off. Carrier A ends up with a significantly lower seat load factor, and, consequently, with a loss. It is not able to win – despite the positive S-curve effect – enough new passengers to

Table 8.3 Effect of capacity addition: carrier A adds one frequency[a]

		Market share	*Pax p.d.*	*Roundtrips p.d.*	*Seats p.d. (one-way)*	*Overall seat load factor*	*Total oper. cost (US$)*	*Total cost per airline (US$)*	*Total ticket revenue*	*Total revenue per airline*	*Oper. profit per A/C (US$)*	*Profit per airline (US$)*
Carrier A	A321	35%	350	1	184	72%	10,120	28,120	10,645	28,000	525	–120
	A320			2	300		18,000		17,355		–645	
Carrier B	B737–500	28%	283	0	0	79%	0	23,400	0	22,640	0	–760
	A319			3	360		23,400		22,644		–760	
Carrier C	B737–800	37%	367	1	170	78%	8,500	25,900	10,620	29,360	2,120	3,460
	B737–700			2	300		17,400		18,740		1,340	
	Total	100%	1,000	9	1,314	76%		77,420		80,000		2,580

[a] Franke, 'Strategic network management'.

sufficiently fill the additional capacity. For carriers B and C, the outcome is at least as negative as for carrier A, because they lose market share, bringing their seat load factors down as well. The overall average load factor on that route slumps from 88 per cent in the equilibrium scenario to 76 per cent – a value that is only slightly above the critical threshold of 75 per cent stated as rule of thumb for continental routes in Chapter 3, section 3.2. Consequently, the cumulated profit on this route goes down by as much as 80 per cent (from US$12,700 to US$2,580). At the same time, the macroeconomic efficiency of the capacity expansion scenario drops from 44 to 9 per cent of the ideal reference scenario.

8.4 Impact of overcapacity in a market

These results show the disastrous effect of excess capacity on the economic performance of air traffic systems – both for single routes and for large geographic areas. Overcapacity is actually one of the most dangerous threats for airlines, as well as for the entire aviation sector. Smart airlines consequently try to avoid significant capacity additions without respective market growth, though a certain tendency towards overcapacity is immanent to each and any air traffic market. The reason is that there are low entry hurdles, but high exit hurdles for founding airlines and building fleets.

During periods of economic upturn demand for air traffic grows quickly, making it particularly seductive for each airline to buy new aircraft and strive for a fair share of the growth. In that context, it is not too difficult to find investors for the new aircraft; aviation has a positive image as an industry and tends to steadily grow in the long run. Dips and downturns occur – such as 2001, 2003 and 2008/9 – but recovery came quickly every time, bringing the industry back to the long-term growth trajectory.

Once the new aircraft are in operation, they show an amazing longevity, staying in the market for three decades or more. Professional maintenance keeps them up and running, and their high asset value makes it reasonable to operate them as long as possible. Even if major international airlines start replacing their aircraft after 15 to 20 years for efficiency and reputational reasons, these veterans of the air are still good enough for a second or even third life at second-tier airlines and cargo carriers. Furthermore, most airlines apply depreciation periods of 12 to 15 years, so, after that turning point in the life cycle of an aircraft, the financial costs drop, compensating for the increasing maintenance cost and reduced fuel efficiency of the ageing aircraft.

There is one aspect that could change the negative outcome of the capacity expansion scenario shown in Table 8.3: the assumption that frequencies added by one carrier do not simply entice demand away from other carriers but also activate additional demand not tangible in that market so far ('latent' demand). This assumption could cure the negative effect of the overcapacity caused by adding frequencies to a balanced market (equilibrium). This assumption is not overly realistic, however, since additional frequencies first of all attract *business* travellers from other airlines who have to take a flight in any case – not latent *leisure* demand, as in the price war scenario (see below).

8.5 Price war scenario

In the *price war scenario*, carrier A decides to fight for a higher market share by lowering its ticket fares by 10 per cent (i.e., yield drops from US$0.08 to US$0.072 per RPK). For the case study, we assume that travellers on the route between Spoke Town and Network City appreciate this discount, and buy more tickets from carrier A.

- Carrier A can entice 5 per cent away from both carriers B and C (50 passengers each – i.e., 100 passengers p.d. one-way in total).
- On top, carrier A is able to win an additional 5 per cent (50 passengers p.d. by stimulating latent demand – i.e., from clients not having purchased tickets for this route previously).

To accommodate the additional 150 passengers per day (one-way), carrier A needs to add – as in the previous scenario – one roundtrip with its larger aircraft type. In this respect, the 'price war' scenario is not pure, but overlaps with the 'capacity expansion' scenario.

The results of this scenario can be found in Table 8.4: compared to the equilibrium scenario, carrier A is able to increase its client base from 250 to 400 passengers p.d. (one-way), and its market share from 25 to 38 per cent. Fifty passengers out of the additional 150 passengers do not come from carrier B or C but from new demand stirred by the lower fares; the overall market size has thus grown by 5 per cent, from 1,000 to 1,050 passengers.

This is the main reason why this scenario leads to an overall profit that is slightly higher than in the previous ('capacity expansion') scenario – US$3,380 instead of US$2,580. Obviously, the additional demand is just able to compensate for the drop in price. Likewise, the average seat load factor on that route slumps from 88 to 80 per cent, but is nonetheless higher than in the 'capacity expansion only' scenario (76 per cent). Nota bene, this scenario is still highly inefficient from an overall market perspective: with its macroeconomic efficiency factor of only 11.7 per cent (compared to the monopolistic reference case), it is only slightly less inefficient than the previous one.

The slight improvement of the overall profit is, of course, contingent on the assumed price elasticity of demand; in this case, carrier A is able to increase its market share by 13 per cent as a reward for 10 per cent lower fares. In an alternative setting with less elastic demand, the overall profit could easily be lower than in the scenario described in Table 8.3. The other driver for the increment between the scenarios in Tables 8.4 and 8.3 (as well as between those in Tables 8.4 and 8.2) is the reshuffling of the unit cost mix as a result of carrier A's competitive moves. If demand is reallocated from carriers with cheaper operations to carriers with more expensive operations, the overall efficiency

Table 8.4 Effect of price reduction: carrier A lowers fares by 10 per cent and adds one frequency to accommodate the additional demand[a]

		Target market share	*Pax p.d.*	*Roundtrips p.d.*	*Seats p.d. (one-way)*	*Overall seat load factor*	*Total oper. cost (US$)*	*Total cost per airline (US$)*	*Total ticket revenue*	*Total revenue per airline*	*Oper. profit per A/C (US$)*	*Profit per airline (US$)*
Carrier A	A321	38%	400	1	184	83%	1,0120	28,120	10,949	28,800	829	680
	A320			2	300		18,000		17,851		–149	
Carrier B	B737–500	27%	283	0	0	79%	0	23,400	0	22,640	0	–760
	A319			3	360		23.400		22.644		–760	
Carrier C	B737–800	35%	367	1	170	78%	8,500	25,900	10,620	29,360	2,120	3,460
	B737–700			2	300		17,400		18,740		1,340	
	Total	100%	1,050	9	1,314	80%		77,420		80,800		3,380

[a] Franke, 'Strategic network management'.

Table 8.5 Effect of price war: all carriers lower their fares equally by 10 per cent and adjust their capacities to accommodate the changed demand[a]

		Target market share	*Pax p.d.*	*Roundtrips p.d.*	*Seats p.d. (one-way)*	*Overall seat load factor*	*Total oper. cost (US$)*	*Total cost per airline (US$)*	*Total ticket revenue*	*Total revenue per airline*	*Oper. profit per A/C (US$)*	*Profit per airline (US$)*
Carrier A	A321	25%	275	1	184	82%	10,120	19,120	10,908	19,800	788	680
	A320			1	150		9,000		8,892		–108	
Carrier B	B737–500	33%	367	2	320	83%	19,200	27,000	19,217	26,424	17	–576
	A319			1	120		7,800		7,207		–593	
Carrier C	B737–800	42%	458	2	340	93%	17,000	25,700	22,881	32,976	5,881	7,276
	B737–700			1	150		8,700		10,095		1,395	
	Total	100%	1,100	8	1,264	87%		71,820		79,200		7,380

[a] Franke, 'Strategic network management'.

of the traffic system must go down – or up, in the opposite case. This unit cost effect of demand reallocation is coincidental but may have a significant impact on overall profit.

In real life, carriers B and C would rather not accept the considerable drain of demand initiated by carrier A but fight back. If they, for instance, follow carrier A in lowering their fares by 10 per cent, that could end up in a price war; this scenario is illustrated in Table 8.5.

Provided that passengers have not changed their fundamental preferences – leading to the initial equilibrium scenario – and have switched to carrier A simply because of the discount, a price-matching strategy by carriers A and B would lead back to the initial market shares. The upside of this scenario is that the overall market size grows by 10 per cent, since some potential clients appreciate the price drop and become real clients. The downside, on the other hand, lies in the deliberate drop of yield. Furthermore, all three carriers need to reshuffle their capacity allocation to account for the additional demand, so the extra volume does not come for free.

8.6 Return to equilibrium after a price war

All in all, the 'price war' scenario leads to a new equilibrium point, since all clients go for the same strategy. On the other hand, this new equilibrium is neither stable nor efficient. The lack of stability stems from the fact that all three carriers are in a worse position than in the initial equilibrium scenario, with carrier B even making losses in total. Carrier C experiences only minor losses compared to the equilibrium case, which is a coincidental effect, though, since it is able to bring up its seat load factor through reshuffling its capacity allocation, almost compensating for the lower fares.

The overall seat load factor is sustainable and close to what it was beforehand (87 per cent instead of 88 per cent); the 10 per cent yield decrease is a heavy burden for the overall profit, however, which plummets from US$12,900 to US$7,380 (from 45 to 25 per cent in terms of macroeconomic efficiency). Consequently, if no carrier gives up and leaves the market, there is strong pressure for all three carriers to stop the price war and get back to the initial scenario. The only winners of a price war scenario are usually the clients, since they get the same services as before for less money.

The exception from that rule are markets with initially inefficient or even quasi-monopolistic competition, resulting in skimming strategies: if the tickets are overpriced in such markets, some potential passengers will stay at home and simply not buy tickets. A comprehensive reduction of the fare level would then attract many new, price-sensitive clients, boosting revenues if the price elasticity of demand is high enough.

In our case study, higher elasticity could consequently change the break-even point of the price reduction scenario: if the additional demand attracted by an overall price drop of 10 per cent amounted to, for example, 20 per cent instead of 10 per cent, the overall profit could be higher instead of lower than in the initial equilibrium scenario. This phenomenon occurred over the past two decades in many mid-sized markets with limited competition as soon as the first low-cost carrier entered the stage. If the latent demand stimulated by the discounted fares is big enough, this could lead to a win-win-win situation: travellers can fly for less money, but all carriers make more profits due to increased volume on the respective route.

8.7 Conclusion of case study

To sum up, the case study discussed in this chapter is a tiny network management scenario, and the input data are completely made up. It is nevertheless realistic enough to illustrate many of the mechanisms introduced in the previous chapters of this section. This finding underlines a basic paradox of network management: networks are by definition complex structures, and the constituting interdependence of resources makes every calculation complex as well. Nonetheless, there is a broad range of experienced-based correlations, trade-offs and rules of thumb that are valid across most network management problems, making the daily work of network management staff much easier. Ideally, most network management scenarios can be cut into slices, or simplified to a level (at least for an initial evaluation) at which the mechanisms discussed in this case study provide useful insights for the final solution.

Part II

Interplay with adjacent functions

The first part of this book, especially Chapter 4, section 4.5, has shown that network management as a core business function of airlines is strongly interlinked with a variety of adjacent functions. There is an extensive exchange of inputs and outputs between these functions; if the interplay does not work, network management will not be able to perform well. For instance, if the fleet of an airline features an inappropriate mix of aircraft types for a given network scenario, network management cannot fully compensate for this fundamental downside. Likewise, a poor crew planning process can easily dilute the benefits of an efficient network schedule.

In this part, the interplay between network management and its adjacent functions is examined in further detail, as well as the prospects for a closer integration between network management and some or all of the neighbouring functions. Figure 9.1 shows the interactions discussed in this part.

9 Interplay and integration with adjacent functions

Figure 9.1, which is based on Figure 4.5, illustrates the interplay between network management and its adjacent functions, such as fleet and crew planning (referred to in the following sections). This chapter discusses the status quo of the interaction, as well as the perspective for a further integration between these planning functions.

9.1 Interplay with fleet management

In the traditional planning approach for airlines, fleet management and network management form an iterative process loop, with fleet planning being the starting point of the loop. As pointed out in Chapter 4, section 4.3, the fleet is the most capital-intensive and least flexible asset of an airline, so it appears to be a pragmatic approach to determine the capacity of the carrier before all other planning decisions. The rationale for building a fleet is based on the following aspects:

- availability of aircraft types from established OEMs (Boeing, Airbus, Bombardier, Embraer, etc.) with a typical combination of aisle numbers, seat capacity, flight range;
- actual availability of a certain aircraft type at a certain point in time;
- price point for the preferred aircraft type at a certain point in time;
- political and tactical considerations: avoidance of dependence on one major supplier;
- commercial considerations: unit cost position, fuel efficiency, maintainability, depreciation pattern, homogeneity of fleet (commonality effect); and
- new aircraft as replacement for existing, outdated aircraft (i.e., 'roll-over') versus capacity expansion.

9.1.1 Iterative fleet and network development

It needs to be kept in mind that an established carrier is never in the position to freely decide on all the above-mentioned aspects, since it cannot rebuild the fleet from scratch in an ideal composition but needs to incrementally update

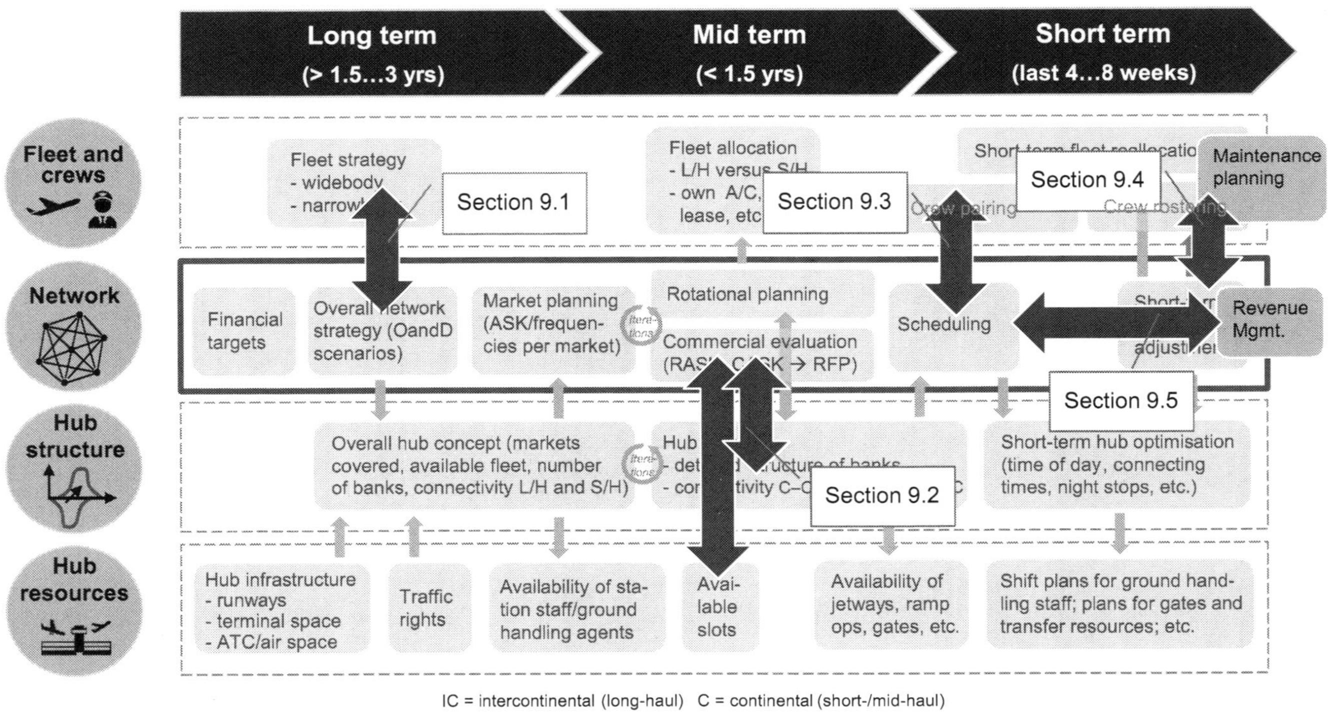

Figure 9.1 Interplay and integration prospects between network management and adjacent functions

an existing fleet from time to time. It may take quite a while until an actual aircraft with optimal operational efficiency and cost position can be reasonably replaced by a newer aircraft type – even if the newer type is more efficient than the older model. Especially if the depreciation period is over after 12 to 15 years, operating this aircraft is often very attractive for up to another ten years, so the exit hurdle is high.

Premium carriers still replace most of their aircraft after 15 to 20 years to sustain their high-quality claim, but most second- and third-tier airlines do not feel such constraints and keep ageing aircraft in service for an additional ten to 15 years. It takes quite a while before maintenance costs and relative fuel consumption go up so much that a carrier with limited investment capacities feels forced to replace a fully depreciated aircraft.

Thus, developing a fleet is always a long-term, permanently rolling process, and a continuous compromise between operating and financing costs. Furthermore, capacity additions – as well as reductions – are very often not driven by the free strategic will of an airline but by external factors, such as economic cycles. During upturn phases the urge is enormous to expand capacities, whereas during downturn phases most airlines try to ground parts of their fleet to achieve relief on the cost side. Since all airlines follow the same cycles, it is usually challenging to buy or lease popular aircraft types during an *upturn*, whereas it is difficult to get rid of underutilised aircraft for a reasonable price during a *downturn*.

Once the basic fleet size and mix is determined, network scenarios are designed to utilise the inherent seat capacity in an efficient way, achieving a profitable match between demand on O&Ds and aircraft capacities on flight legs. It may well occur that the network scenario and respective production portfolio evolve and diverge from the situation initially taken into consideration for the previous fleet plan, so the assumptions for the next iteration loop may have to be adjusted accordingly.

In a worst-case scenario, it may turn out during the network design and scheduling process that the available fleet is not very well suited to operate the envisaged network scenario, for instance in terms of aircraft size (i.e., seat capacity), range, unit cost position or maintenance requirements. Unless fleet management has done a very bad job in the past, this scenario can occur only if the envisaged network structure represents a disruptive change compared to the previous structure – which is a rare case.

If it occurs, however, the leeway for the respective carrier to massively change the structure of its fleet is very limited – at least, at short notice. Network and fleet managers would have to muddle through with the existing fleet, accepting certain inefficiencies because of the mismatch between fleet structure and network structure, and initiate a stepwise roll-over programme for the upcoming years – provided that the required capital for this investment is available. In all other scenarios, incremental adjustments of the fleet should be sufficient to cover incremental changes of the network structure.

9.1.2 Integrated fleet and network development

The main downside of separate fleet and network planning is the permanent latent mismatch between the fleet ideally needed for a certain flight programme and the actual fleet in place. The inherent loss of efficiency may be small for only limited changes of the network structure, but it is always there, and it dilutes value.

For an integrated planning approach, on the other hand, there is one major challenge: the different reaction times for adjusting fleet and network structures. Whereas a certain subset of a network, such as a route, can be changed relatively quickly – provided that required slots and local handling resources are available – it is rather difficult to get rid of an aircraft at short notice, and replace it by a new one.

The radical solution for this dilemma is the introduction of a fully homogeneous fleet, with only one single aircraft type. This is the paradigm pursued by low-cost carriers, since it not only simplifies the planning and allocation process but also unleashes highly welcome cost reduction potentials. A homogeneous fleet reduces complexity costs in aircraft maintenance, crew training and licensing and on-board processes. This approach reduces the interplay between fleet and network development to the simple adjustment of aircraft numbers: if the new network structure requires more capacity – in terms of ASK – than the current fleet is able to deliver, the next aircraft of the same type needs to be purchased.

Unfortunately, this solution works only for a homogeneous flight programme with a small range of market sizes and stage lengths, making aircraft basically exchangeable. For hub and spoke carriers, this approach is consequently not viable; the split into short-/mid-haul and long-haul routes requires fundamentally different types of aircraft (narrow-body versus wide-body aircraft), and even within the two clusters there are routes with very different characteristics. A small market with a high share of high-value point-to-point demand may well justify a relatively small aircraft, whereas a trunk route with intensive competition and many transfer passengers on board can be served in an economically sustainable way only by deploying a large aircraft type. In the former case, the non-stop service itself is sufficiently differentiating to raise optimal fares, whereas in the latter scenario the only success factor is a competitive unit cost position.

Furthermore, hub and spoke carriers need a certain flexibility in terms of capacity size even on single routes, since they may have to react to moves by competitors. As described in Chapter 6, section 6.3, Chapter 7, section 7.1, and Chapter 8, players in a certain market tend to observe the S-curve effect, looking for the most profitable balance between frequency number and aircraft size. If one carrier increases its capacity to gain more market share, the other airlines active in that market may feel forced to match this move. If they decide to do so, they will add frequencies, and deploy smaller aircraft types to keep seat load factors in a sustainable range. This response is, of course, contingent on the availability of different aircraft types; an airline with a one-aircraft-type strategy would be unable to react to such market events.

In practice, hub and spoke carriers need to find the sweet spot between a fully homogeneous fleet – with its immanent complexity cost advantage – and a very versatile fleet composition, enabling the airline to respond to each and every market challenge. As a rule of thumb, a mid-sized carrier needs at least two wide-body aircraft types, and two to three narrow-body aircraft types. Having fewer than that is favourable in terms of complexity costs but limits the ability of the carrier to adapt its fleet to internal or external changes. More than that adds flexibility, but comes at a price. Both deviations from the rule of thumb may be well justified, if a certain airline has a specific business environment.

A fully integrated planning approach would require a new kind of optimisation model, with a simultaneous allocation of resources on three levels:

- medium/long term: design of fleet and network scenarios
- medium term: elaboration of flight schedule (market plan plus rotational plan)
- short term: adjustment of capacity allocation and rotational plan

The coincidence of fleet and network design is easy to achieve if aircraft are purchased for and dedicated to certain routes. Many US carriers pursue this practice, at least for long-haul services. In this case, there is a direct link between aircraft and route, and a simultaneous planning or optimisation process is intuitive.

For carriers with more complex rotational patterns and allocation rules, however, leading to permanent changes in the match between aircraft and route, this natural coincidence does not occur. This is the most common scenario for major airlines on short- and mid-haul routes: it is by no means an advantage to have a fixed allocation between a specific aircraft on a tail sign level and a certain route; on the contrary, it is beneficial to maintain a considerable flexibility in assigning aircraft (tail signs) to routes. It is not necessary – and not even helpful – if a certain route that has been served by a specific aircraft yesterday will be served by exactly the same aircraft today or tomorrow – at least, not in scheduling.

Once the flight schedule is designed, flexibility becomes less helpful, and a certain stability in the allocation of aircraft to routes and flights is desired. A specific aircraft entails a variety of operational requirements, such as aircraft slots, cockpit and cabin crews with the respective licences, and maintenance capacities; switching an aircraft from one route to another means adjusting the operational requirements as well.

Some airlines have nonetheless tested short-term changes under the label 'dynamic fleet reassignment'; this method was based on the assumption that a deliberate fluctuation in the match between aircraft (type) and route generated more upsides than downsides. The upsides were short-term improvements of seat load factor in cases of fluctuating demand, whereas the downsides were the cost of flexibility, especially the cost for reshuffling schedules – including all operational requirements – again and again. In its purest form, the 'dynamic fleet reassignment' approach would have regularly reshuffled aircraft allocations at short notice even if external conditions such as forecast demand for a certain route had not at all changed since the last update.

In practice, this highly demanding approach – both in terms of mathematical model and process – has not proved to be valuable enough to make its way into the standard network management toolbox of hub and spoke carriers. The effort of reshuffling the fleet at short notice was too high to be compensated for by the extra efficiency gains. Nota bene, every airline will reshuffle its fleet allocation at short notice if external trends make all previous scheduling assumptions obsolete, and cause unsustainable seat load factors; this is an exceptional approach to deal with crises, though, and not a standard procedure.

If the coincidence between fleet and network design is difficult to achieve, there is another way to establish a more integrated planning process between the two of them: a model that does not accept aircraft as being expensive and difficult to purchase but treats them like any other resource needed to realise a certain network scenario. If the planners decide for the most favourable integrated scenario, they assume that the respective capacity can be provided – at reasonable cost.

The respective scenarios could become so complex that the traditional linear programming algorithms (cf. Chapter 5) must be updated, or even replaced by other resource allocation models. Examples for advanced applications are *machine learning* (ML) or *artificial intelligence* (AI) models, which are in vogue these days (see Chapter 12). Machine learning models are not by nature smarter than traditional mathematical models, nor can they simply replace LP solvers. They can be trained to copy human behaviour and automate it, though. In this concrete case, a meaningful application of an ML model could be the build-up of efficient network scenarios – a task that is still heavily based on human experience, and less on mathematical models. Network planners use their experience and expertise about efficient scenarios as a starting point for the next iteration, which is then calculated by an LP solver.

9.2 Interplay with hub management

9.2.1 Sequential network and hub planning

In Chapter 6, section 6.5, the traditional approach of hub and spoke to hub planning – in correlation with network planning – has been explained. It is based on a sequential process featuring the following steps:

- definition of *financial* and *strategic targets* for the *airline*;
- definition of *financial* and *strategic targets* for the *airline's network* (offered destinations, market coverage, market shares, passenger numbers, ASKs, customer segments, fare mix, ticket revenues, network profit, etc.);

- development of a *fleet plan* suited to accommodate the business volume envisaged in the network strategy (passengers and ASKs on certain O&D bundles or routes);
- design of *network scenario(s)* deploying the available/planned fleet in an efficient way to process the business volume, and cover the destinations/markets/customer segments envisaged in the network strategy;
- design of commercially viable flight schedules (market plans);
- elaboration of operationally viable flight schedules (rotations);
- definition of a hub structure (i.e., number and shape of banks) suited to build the connections required for the flight schedule; and
- allocation of the required resources to operate the hub.

This process is strictly top-down, so it always starts with the strategic perspective, followed by the market, aircraft roundtrip and hub view. This sequence is based on the paradigm that the network is designed in a way that ensures the most efficient allocation of resources, and generates maximum profit, whereas the hub should act as a mere production platform, delivering whatever it takes to make the network vision happen. The underlying assumption is that the hub management can basically realise each and any traffic pattern that the network management has defined for that hub – including the traffic volume, connections and rotational patterns constituted by the respective network scenario.

On a detailed level, there may well be iterative adjustments of the network initiated by the hub management function; for example, if the hub management realises that preponing or postponing a certain flight for, say, ten minutes could considerably improve the operational stability or efficiency of the hub, network planners may check that change and move the respective flight, provided that:

- the required new slot is available;
- all other resources needed to operate this flight are available earlier (or later) as well;
- all envisaged connections for passengers on that flight can be maintained; and
- preponing or postponing the flight does not deteriorate its relative attractiveness compared to competing services.

This case is rather rare, however. It is much more common that network planners decide to bring forward or postpone a certain flight in a hub, for the sake of an improved connectivity, or a more attractive timing compared to relevant competitors serving the same route. Network planners often assume that their colleagues from hub management can handle basically every conceivable change of the schedule – which, of course, is not true in real life. Nonetheless, the traditional top-down planning sequence is still in place: strategy determines fleet, fleet determines market plan, market plan determines operational plan (rotations), and operational plan determines hub plan (cf. Figure 6.4).

9.2.2 Integrated network and hub planning

The assumptions from the previous paragraph hold sufficiently true for all scenarios with stable operational environment and limited changes. The top-down approach does *not work properly* if the changes of the hub structure are not incremental, but fundamental. Reasons for a fundamental rethinking of the hub could be the following:

- the overall traffic volume of the home carrier in his hub changes – e.g., as a result of economic upturns or downturns;
- the number of banks is too large or too small for the current traffic volume;
- the position and shape of the banks does not support the connectivity required to realise the network strategy and flight schedule;
- the fleet operated by the home carrier in the hub does not support the commercial factors assumed in the network strategy and flight schedule (e.g., size of deployed aircraft impacting economies of scale and unit cost position);
- the operational constraints of the hub change (e.g., operating hours/night curfew, movement capacities);
- the ratio between point-to-point demand (from the local catchment) and O&D demand (from transfer passengers) does not support enough high-value long-haul traffic; and
- the broad coverage of destinations from the hub ('360 degree approach') needs to be reconsidered – e.g., because of structurally unprofitable feeder de-feeder services.

If one or more of these reasons occur, the match between the hub (bank) structure and the network pattern may get lost, resulting in a deterioration of the hub's performance. The hub management loses its ability to adapt hub operations to the requirements of the network concept – which is a fundamental assumption behind the traditional top-down planning paradigm. If this assumption does not hold true any more, the respective planning approach may lead to dysfunctional hub operations.

For example, if the number of banks has been chosen to establish efficient rotations with a good productivity level, the overall performance of the hub will be satisfactory only if each bank is sufficiently filled with inbound or outbound flights; in this case, both asset utilisation and connectivity achieve high values. If, however, the banks are half empty, because the actual number of flights had to be stretched too thinly across the banks, the connectivity will go down, not up, spoiling the commercial result of the hub (see second and third bullet points in the list above).

Likewise, the ratio between point-to-point and O&D demand operated via the hub is a fundamental KPI of a hub. Hubs with a small share of point-to-point demand show a tendency to unprofitable traffic patterns. Although a hub is, first and foremost, a production platform – and thus a cost centre – the traffic mix available in the hub has a significant impact on the profit the hub carrier can achieve with its network. The fare level for point-to-point demand is usually higher than that for O&D demand, since the non-stop service is always the most popular offering for a client. If a traveller has the choice of a direct flight for his personal O&D, he will book a connected flight on that O&D only if he is granted a considerable discount from the respective carrier.

Consequently, an airline may be forced by external or internal conditions to reshuffle its hub structure and operational model, since sticking to the initial structure would dilute value. The problem is that most airlines have neither experience with nor planning methods for a bottom-up reshuffling of their hub. Changing the hub structure beyond incremental adjustments is thus an immense risk for every airline, since the balance between network and hub structure may be an unstable one: moving fundamental operational parameters of the hub such as number or shape or banks may, in the worst case, entail a massive worsening of the network performance.

As already stated in Chapter 6, section 6.5, and Chapter 7, section 7.1, the solution for this dilemma is to perceive the bank structure of a hub not only as *a result of* but also as *an input for* fleet and network scenarios. Some hub carriers have already tested reverse planning models to close the gap and better understand the impact of new hub operating models on the performance of the network. Such models permit the calculation of different bank structure scenarios not only with respect to hub performance indicators (connectivity and productivity of resources) but also related to the effectiveness of rotations, the coverage of markets, the profitability of network structures and – ultimately – the effectiveness of a given fleet.

This reverse perspective on hub and network structures is not really an integrated approach in a narrower sense, but it is at least bidirectional (top-down *and* bottom-up) and puts network and hub structures on the same analytical level. Besides validating a new hub structure or operating model – e.g., a reduction of the number of banks, or the extension of a night curfew – this new approach has another beneficial side effect. The conventional top-down approach leads to a practice in which the hub structure is adapted to the given fleet and network concept of a carrier just once – at the very beginning. The efficiency of this initial hub concept is usually mediocre, so hub planners improve the performance through continuous incremental optimisation steps, such as adjusting the shape and timing of banks. Slight shifts of inbound or outbound flights – as small as a few minutes only – may add valuable percentage points in terms of hub productivity or network profitability.

For a while – even up to several years – incremental changes of the network and incremental adjustments of the hub model may work quite well together. Sooner or later, however, this path is a dead end: although a network can fluctuate forever – assuming that demand and capacities are always volatile – the fine-tuning of a given hub structure within a fixed operational environment will ultimately converge towards an ideal status that can no longer be improved any more. Once the optimal stage is reached, only a major reshuffling of the hub concept (e.g., of the bank structure) is strong enough to break up the stalemate situation.

Even worse, the dead end situation at the end of an incremental hub improvement cycle may not only prevent any further adjustment of the hub structure – and thus paralyse the planning process – but, in the worst case, lead to a highly unstable scenario in which the level of fine-tuning is so delicate that the operational stability gets lost. One single flight delay may trigger a domino effect and impact all subsequent flights, until a major share of all flights touching the hub are impacted. From this critical status, there is almost no escape, because every further change will either dilute the delicate efficiency or damage the operational stability.

With the bidirectional approach, the interplay between network and hub structure can be validated from time to time, and, once a dead end situation is foreseen, alternative hub scenarios can be tested and evaluated with respect to their feasibility and impact on the functioning of the respective network.

9.3 Interplay with crew planning

9.3.1 Iterative network and crew planning

As already pointed out in Chapter 4, subsection 4.3, current network planning methods unravel the planning process by network asset or resource. In other words, the most expensive – or least flexible – resource is determined first, after which all other resources follow in the sequence of declining capital lock-up or cost. In line with this rationale, aircraft rotations are planned first, followed by crew rotations. A separation of the two planning processes is intuitive, since crew roundtrips are different from aircraft roundtrips, and crew limitations are very different from aircraft limitations.

Chapter 6, section 6.4, has introduced the two-step approach used by most airlines: *pairing* is the process of allocating a virtual crew to the different flight segments of an aircraft roundtrip, until the aircraft is sufficiently staffed. This virtual crew has the required licences and type ratings to cover all flight segments in all aircraft types on their roundtrip, and all standard limitations such as maximum flight hours are observed. In total, all virtual crews summed up across the network for a certain point in time do not exceed the total crew body – cockpit and cabin – of the airline. The virtual crews do not yet contain real people, though. This link is created in the second step, the *rostering* process. Four to two weeks prior to a monthly flight period, the paired roundtrips are populated with physical shifts of real crew members.

Pairing and rostering can, depending on the size of the airline, be very complex allocation and optimisation problems. Thus, most major airlines use crew planning systems from specialised IT providers, such as Jeppesen or Lufthansa Systems, to support their crew-planning staff. These tools contain database functions for managing all available crew members, rule-based allocation mechanisms, optimisation capabilities and a function for generating individual shift plans.

9.3.2 Integrated network and crew planning

To simultaneously plan aircraft and crew rotations, a complex equation system must be solved, since crew roundtrips may overlap with several aircraft roundtrips, and vice versa. The equation system alone could be easily handled with a standard Gauss solver (a method invented by German mathematician Carl Friedrich Gauß around 1800, transforming the coefficients of a linear equation system into a so-called row echelon or triangle form). The real challenge comes from the constraints of the allocation problem, though, caused by the long list of contractual or regulatory limits to be observed when allocating cockpit and cabin crews. These limitations and contractual restrictions may vary from one flight attendant or pilot to another, creating an incredibly complex matrix of constraints that is almost impossible to be translated into a mathematical model – at least, into traditional deterministic models.

It is difficult to prove that an integrated optimisation approach for both aircraft and crew rotations is, first, superior to the traditional sequenced method and, second, worthwhile. Since crews are much 'smaller' elements of a network than an aircraft, and much less expensive, it seems to be somewhat overengineered to strive for an integrated approach just for the sake of a more elegant solution. On the other hand, the complex constraints in any case represent a modelling challenge, and solving this issue could quite probably be rewarding.

As stated above, the main limitation of traditional solvers is that every single regulatory or contractual rule needs to be accurately described as a constraint of the core target function (aiming at maximum available capacity on defined flight segments at minimum cost). A methodology that could master this challenge is *machine learning*, or its derivative *deep learning* (also known as artificial intelligence: see Chapter 12).

If enough data points are available, a machine learning model can be trained to tackle a complex cognitive problem like a human being with perfect expertise and overview. For instance, machine learning algorithms are able to learn human languages, understand the inherent concepts and respond in an appropriate way. Being trained with relevant occurrences of the cognitive problem, the model keeps learning by continuously building links between information items in the form of a *neural network* – a technique very similar to the way a human brain digests new information.

For the aircraft and crew allocation problem, a machine learning model could learn which combination of aircraft, crews and restrictions leads to efficient results, and which does not. Nota bene, the machine learning model *is by no means* smarter than a traditional deterministic model in solving the complex equation system of target function and constraints, but it *is* much better prepared to understand the impact of dozens, if not hundreds of parameters on the efficiency of a solution. In fact, one of the strengths of machine learning algorithms is that they can be trained to detect correlations between hundreds of parameters when every statistical method would inevitably fail. Opposed to the traditional models, for which the optimum mix of parameters is an *explicit* result of the relevant equations, the machine learning model identifies efficient solutions based on an *implicit* understanding of the drivers.

9.4 Interplay with maintenance planning

9.4.1 Iterative network and maintenance planning

Similar to the crew resource discussed in the previous chapter, aircraft maintenance capacity seems to be a non-critical resource at first glance, being much more flexible and much easier to allocate to a flight schedule than aircraft, airspace or airport infrastructure. Many major carriers operate their own MRO (maintenance, repair, overhaul) premises, offering their services to their own airline (captive client) as well as to third-party, non-captive clients.

In addition, there are a variety of independent MRO service providers around the globe for light and heavy maintenance. This perspective is one-sided, however, if not short-sighted. It is in fact not very difficult to find the required MRO capability and capacity once a certain aircraft needs to undergo a specific maintenance event. Nevertheless, it is by no means certain that this capacity is available in the perfect location with the perfect timing – let alone with a compelling value-for-money ratio.

Before further analysing the interplay between network management and maintenance planning, a quick wrap-up of maintenance activities may be helpful. First, there is a significant difference between planned and unplanned MRO events. 'Unplanned' means that a vital part of an aircraft fails or breaks down, so that the aircraft cannot fly any more (causing an AOG event: aircraft on ground). The broken part needs repair or replacement – in most cases at the airport where the failure occurred. This causes not only a direct cost – since both a technician certified for this aircraft type and part may need to travel to this airport, and a spare part has to be expedited there as well if the original part cannot be repaired – but also an indirect irregularity cost, since the grounded aircraft causes a delay in the respective rotation if no replacement is available.

Consequently, airlines do everything they can to avoid unplanned maintenance events. The first lever is a rigid regime of planned maintenance events, as explained below; the main idea behind this approach is to regularly replace specific parts of an aircraft *before* they can fail at all. The second, more advanced lever is called *predictive maintenance*, a new approach combining the power of extensive telemetric surveillance with new forecasting methodologies. Hundreds of sensors mounted all over the aircraft continuously scan core parts and functions of an aircraft, such as jet engine fan blades, avionics, hydraulic pipes and structural parts of the hull, sending thousands of telemetric data per minute to the data centres of airlines, aircraft manufacturers and engine manufacturers. Today these data are used ex post to understand the systemic reliability of aircraft components, and to analyse occurring failures.

In the future, these data can be digested in real time by advanced machine learning or deep learning models, learning to detect complex patterns hinting at a potential future failure (e.g., the coincidence of certain temperature and pressure values in the first stage of a jet engine turbine). As soon as the predictive approach is reliable enough, the current model of prescribed maintenance events (see below) could be extended or even replaced by predictive models, suggesting maintenance activities in real time whenever it seems to be required. It will probably still take some time for this advanced maintenance practice to become fully effective, however, since not only aircraft and engine OEMs (original equipment manufacturers) need to approve it but also all international and national aviation authorities.

Opposed to predictive maintenance, prescribed MRO activities are not performed when they seem to be necessary, but when design engineers of the OEMs and aviation authorities deem them appropriate and sufficient to minimise the statistical risk of a failure. Maintenance events are determined by two drivers:

- the number of 'cycles' (i.e., take-offs and landings), because each cycle includes particularly stressful elements for an aircraft (e.g., pressurisation of cabin, full engine power for take-off, mechanical shock of landing); and
- the number of flight hours, because this determines the overall level of usage inflicted on an aircraft.

For short-haul flights, the dominating driver for ageing is the number of cycles, whereas it is the number of flight hours for long-haul flights. The respective maintenance event needs to be initiated – within a certain period of time – if either of the two criterion thresholds is reached (whatever occurs first).

9.4.2 Standard maintenance events: 'letter checks'

Airlines are, basically, entitled to define their very individual maintenance regime for their fleet and operational pattern, as long as the respective OEMs and aviation authorities approve this plan. Since the approval of a customised maintenance plan is time-consuming, though, most airlines still base their plans on the recommendations of the OEMs, consisting more or less of the following traditional maintenance events ('letter checks').

- *Daily check*: as the name suggests, this is the most frequent check, performed every day by a technician. The technicians will walk around the aircraft and through the cabin, checking the integrity of vital components as well as the status of consumables such as lubricants. This daily check is complemented by the 'exterior inspection', or 'walk-around', that a pilot needs to perform prior to a every single flight.

- *A-check*: occurs every 300 to 600 flight hours (i.e., every six to eight weeks); requires 50 to 80 working hours, which are accomplished overnight if possible; includes more intensive scanning than daily check, plus a wide range of functional checks and smaller repairs.
- *B-check*: historically designed for Boeing aircraft, but almost entirely replaced by several A-checks; today applied only for Boeing's B737–200 and B747–200; occurs every 1,000 flight hours (every three to four months); requires around 150 working hours and a downtime of at least 12 hours.
- *C-check:* occurs every 1,500 to 2,000 cycles (i.e., every 18 months to two years); requires around 2,000 to 5,000 working hours over one to two weeks; very intensive screening of all structural parts of the aircraft (fuselage, wings, nacelle, aileron, empennage).
- *IL-check* (intermediate layover): occurs every four years; deeper screening of structural parts; assembly of technical upgrades offered by the OEM; refurbishment of cabin interior.
- *D-check:* occurs every six to ten years; requires 30,000 to 50,000 working hours over four to eight weeks; complete knockdown, overhaul and re-assembly of all major aircraft components; removal of external painting and application of new paint.

The mentioned activities also differ by location: daily checks can be done where the aircraft stands (e.g., at its ramp or parking position on the apron of an airport), the A- and B-checks are usually performed in standard hangars, and the C- and D-checks require so much space and special tooling that they are often conducted in aircraft assembly hangars. The location is also a question of time consumption and cost: it is common to bring technicians and material *to the aircraft* for daily checks and A-checks – because that reduces the downtime of this expensive asset – whereas the aircraft may be transferred *to the MRO premises* for C- and D-checks – including third-party offshore facilities. The latter events are so comprehensive that the required transfer flights cause only a minor increase – in relative terms – of the overall downtime. This rationale appears also reasonable from a cost standpoint: an A-check costs thousands of US dollars, a transfer flight tens of thousands, and a D-check millions; consequently, a transfer flight should be avoided for an A-check, but can be afforded for a C-, IL- or D-check.

Regulatory bodies and airline professionals use an additional distinction for maintenance events:

- *line maintenance*: activities that can be performed 'on the line' – i.e., during the normal ground time of an aircraft rotation (daily check, A-check, B-check); or
- *base maintenance*: activities that can be performed only in the home base of the airline, or of a specialised MRO service provider (C-check, IL-check, D-check).

Alternatively, MRO manuals use the term 'heavy maintenance'. The 'Aviation glossary' defines heavy maintenance as 'any maintenance, inspection, repair, or alteration a manufacturer has designated that requires specialized training, equipment, or facilities'.[1] This definition is valid for most work packages of C-, IL- and D-checks, so most practitioners use the term 'heavy maintenance' as a synonym for 'base maintenance'. 'Base maintenance' refers to the premises required for a certain maintenance event (e.g., for a 'letter check'), however, whereas 'heavy maintenance' points to the licences required for certain maintenance activities.

9.4.3 Integrated network and maintenance planning

Although access to maintenance capacities seems to be easy, maintenance is still a mission-critical function for the flawless operations of an airline's network. Planned ground time is already costly, since it prevents the aircraft from earning money, but unplanned down time (i.e., AOG) is even worse. Furthermore, base maintenance events are by all means expensive, so the timing and frequency of these events have an impact on the economic result of an airline. The approved maintenance plan defines a time window for a certain check of an individual aircraft, so there is some leeway for the airline to ground the aircraft (and hand it over to MRO operations) sooner or later.

If an airline tries to simply minimise the lifetime maintenance cost of its fleet, it should always push the handover of its aircraft for a certain check to the very last day of the respective time window determined in the maintenance plan, because this approach ensures the minimal overall number of maintenance events. This approach can still be suboptimal, however, because the latest date for grounding an aircraft is not necessarily the most efficient date.

For instance, an aircraft needs a replacement during its grounding; it is conceivable that the replacement can be realised in the rotational plan at lower cost at an earlier date in the maintenance time window. Alternatively, there could occur a drop in demand at the beginning of the maintenance time window, so bringing the respective maintenance slot forward may help the airline in downscaling its capacity at short notice and, simultaneously, making best use of the idle time.

These examples raise the question of which entity should have the final call with respect to the decision on timing and frequency of aircraft checks: the airline, or the maintenance service provider? Since it is the airline that is accountable and liable for the compliance of its maintenance plan with official regulations such as EASA CAMO Part M, the answer seems to be easy. From a commercial standpoint, however, finding the sweet spot for both the aircraft rotation and the utilisation of the maintenance premises could add considerable value.

This is particularly relevant if the maintenance entity is not an external supplier but part of the same corporation as the airline. Some airlines have tested both options – the airline *or* the maintenance entity having the final call for maintenance planning – but there seems to be no clearly superior option.

A recommendable compromise may be a joint task force managing the interface between airline and maintenance operations, taking into account the requirements of both parties.

The advanced approach of 'predictive maintenance' mentioned earlier in this chapter would be one way to establish integrated network and maintenance planning. With permanent information on the state of vital components, the decision on maintenance activities would become a daily, real-time task, turning maintenance into an integral part of every aircraft rotation (similar to the current daily check), instead of a special event outside the regular flight operations. Nevertheless, it has already been conceded that it will be a lengthy process until predictive maintenance as a methodology receives the necessary approvals to replace the traditional system of predefined 'letter checks' with clear trigger values (number of cycles, number of flight hours).

9.5 Interplay with revenue management

9.5.1 Sequential network and revenue management

Like network management, revenue management is one of the most success-critical functions with effect on an airline's revenue. Figure 4.6 has illustrated the interplay between these two functions. The third key commercial function – not highlighted in Figure 4.6 – is *pricing*, which also has close interfaces with the other two. Together, these three tasks form the commercial triangle of an airline, ultimately providing the distribution and sales staff with the information they need to sell as many tickets as possible at the optimum fare:

- offered services (destinations, flights, timing, product features)
- segmentation of offering (i.e., booking classes)
- availability of booking classes for a certain flight at a certain point in time for a certain price
- terms and conditions for these prices

Network management has a long- and mid-term focus (with short-term adjustments if required), but revenue management is clearly a short-term activity. What both functions have in common is that they reconcile commercial (demand) with operational (capacity) aspects. In a simplified way, it could be stated that *network management* allocates a *given capacity* (available in a certain format, such as aircraft types on routes and roundtrips) to *expected demand* for air transportation (on O&Ds), whereas revenue management allocates *occurring demand* (for tickets on certain routes) to available capacity (in the form of seats on suitable flights). This *double allocation* or *dual matching* process is a key success factor of airline management.

Another differentiation is the measurement of success: as explained in Chapter 3, section 3.3 and Figure 3.1, the design and optimisation of networks exerts a decisive influence on the profit of an airline, depending on the remaining time till a certain flight event, which is accounted for in the common profit contribution scheme. At every stage of this scheme, network management decisions have an influence on both revenues *and* costs – unlimited in the long run (profit contribution level 3: PC3), and only incremental in the short run (profit contribution level 1: PC1).

Opposed to this, revenue management has a smaller manoeuvring space, with most of the capacity decisions – and, thus, costs – already determined. Consequently, most airlines use the simplified but pragmatic assumption that all relevant costs are completely fixed at this late stage before the flight event, so that maximising the *profit* means maximising the *revenue* for a given capacity on a certain flight – a concept that is, obviously, reflected in the name 'revenue management'.

To accomplish this task, traditional revenue management concepts are based on five pillars.

(1) *The segmentation of capacity*: to extract the maximum value from the price/demand curve (*'consumer surplus'*), a supplier should break down its inventory into segments, addressing customers with different combinations of readiness to pay and expected service level. The larger the number of segments, the more consumer surplus the supplier will be able to extract from a certain market. In theory, the ideal case would imply an infinite number of segments, extracting 100 per cent of the available consumer surplus. Segments need to be maintained, however (which absorbs work time and management attention), so that, in practice, the number of segments is limited, if not small. At airlines, segments are called 'booking classes' or – in internal conversations only – 'buckets'.

(2) *The forecasting of demand per segment*: as with network management, the forecasting of demand is a mission-critical capability. For revenue management it is even more complicated, since predictions are needed not only for O&Ds or single flights but for each booking class on this flight. For example, if a flight is likely to be ultimately fully booked, the acceptance of ticket requests needs to be performed with the utmost care, to avoid a dilution of potential revenue. Why sell one of the last remaining seats to a cheap booking class, if the forecast claims that there will still be demand for more expensive booking classes? If that situation occurs, the cheaper booking classes will be closed by revenue management, leaving only the high-yield booking classes open. Vice versa, it is useless to offer tickets in expensive booking classes if the seat load factor of a certain flight is likely to be poor, and the remaining demand will be price-sensitive. In this case, the cheaper booking classes should be wide open, at the expense of the more expensive classes.

(3) *Fencing*: in traditional revenue management concepts, the underlying assumption is that booking classes must be fenced – i.e., differentiated by service levels or booking rules accepted by the clients. The most common fences are: (i) cabin compartment/class (first class, business class, premium economy class, economy class). On long-haul flights in particular, cabin

classes are clearly associated with different types of seats, comfort features and service levels (ii) pre-booking discounts (the earlier a passenger books his flight, the more likely it is that he gets access to cheaper booking classes) (iii) time-of-day or stay-related restrictions (on the assumption that business travellers are more time-sensitive and less flexible in their travel plans than leisure travellers, the latter can get access to cheaper fares if they accept restrictions that the former would find prohibitive, such as overnight or weekend stays before the return flight).

(4) *Nesting*: to be consistent, every booking class needs to contain not only its own capacity (allocated to this booking class as a subset of the overall inventory for a certain flight, based on the forecast for demand and readiness to pay per segment) but also that of all cheaper booking classes (the 'nesting' approach, resembling the effect of the Russian 'matryoshka' dolls). If high-yield demand is stronger than forecast for a certain flight, it is not necessary to close cheaper booking classes and reallocate the respective capacity to more expensive booking classes; the revenue management system will automatically grant access to all cheaper booking classes in this case.

(5) *Overbooking*: it is a common phenomenon in air travel that clients having booked a flight do not show up once the flight gets operated ('no-shows'). Business travellers in particular are frequently forced to reshuffle their travel plans – a need that is reflected in one of the typical benefits of business class tickets: the right to rebook or cancel a booked flight at any time without extra cost. For those travellers – business or leisure – who need this flexibility but cannot afford an expensive business-class tariff, a full-fare economy-class ticket usually grants the same right. In any case, no-shows make the work of revenue managers much more complicated, because they turn the balancing of incoming demand with available capacity into a hunt for a moving target. Even if a flight appears to be fully booked shortly before the flight event, no-shows can push the ultimate seat load factor down to below 100 per cent, diluting revenue potential that would have been earned if the load factor could have been stabilised at 100 per cent. The remedy for this spillage effect is so-called *overbooking*: airlines track the no-show rates for the available booking classes on certain flights and allow booking rates of more than 100 per cent to compensate for the no-shows without losing revenue potential. If, for example, a certain flight featured an overall no-show rate of 4 per cent throughout the past couple of flight events, raising the limit for bookings to 104 per cent of the physically available capacity during the booking period should, ideally, lead to an ultimate seat load factor of 100 per cent. Forecasting no-show rates will never be perfect, though, so the actual result may be suboptimal: if the actual no-show rate happens to be higher than expected on a certain date, some revenue is lost, whereas, if it is lower than expected, the number of passengers exceeds the number of seats. Airlines try to solve the latter problem by offering a rebooking plus a premium to passengers who are presumed to be less time-sensitive than others, such as students.

This approach is called 'voluntarily denied boarding', and it works in the majority of all respective cases.

It becomes obvious from these five pillars that revenue management is a highly analytical function, depending in its performance on the availability of accurate forecasts and smart algorithms. This is another similarity with network management: both functions are based on complex models and econometric data, just with different time horizons. Since the costs are mostly fixed shortly before the flight event, revenue management can be said to be successful if it achieves at least the load factors and yields assumed by network management in the scheduling process; overperformance is, of course, highly welcome.

The success of revenue management is strongly biased by the current market environment, however: in a favourable scenario, mediocre revenue management decisions can still lead to results *above plan*, whereas a weak market is likely to cancel out even excellent performance on the part of revenue management, and end up with revenues *below plan*. It is well-nigh a characteristic of revenue management that its success is almost impossible to measure, as the influence of demand volatility and competition is too high. On the other hand, various statistical studies on KPIs such as yield and load factor have provided solid evidence for the estimate that the application of state-of-the-art revenue management practices leads to a revenue increase of – on average – 2 to 3 per cent compared to a base case without revenue management activities.

Unfortunately, as stated above, this calculation cannot be made for a single flight. In any case, this overall effect is sizeable in an industry in which only a few airlines are able to achieve more than the average operating margin (i.e., EBIT) of 5 to 8 per cent in good times (let alone the losses in times of downturn), keeping in mind that the additional revenue almost entirely goes into the bottom line of an airline.

In fact, the appreciation for the positive effects of revenue management was so high that, soon after its introduction in the first half of the 1990s, it took over a crucial role in the commercial steering of an airline, in some ways even displacing the *pricing* function. Like network and revenue management, pricing and revenue management are adjacent and closely related functions. Whereas pricing tries to find the optimum fare for a certain offering *in the medium term* – given a certain affinity to pay of customers and a certain competitive environment – revenue management aims at segmenting the offering in a way that ensures the maximum extraction of customer surplus (and, thus, revenue) shortly before the flight event. There are of course short-term activities of pricing as well, such as promotions per customer segment, channel and geography, but, overall, the short-term optimisation of revenue has become the domain of revenue management.

Although pricing is still needed to reflect market effects such as the elasticity of demand or competition, the actual price that a client ultimately pays for a certain service level (i.e., booking class) on a certain route is mostly determined by revenue management: if demand increases or decreases on a certain O&D, the airline is not directly forced to raise or lower its prices, but can simply open or close respective booking classes. This approach is much smoother and less risky than adjusting prices, especially than lowering a price in times of poor demand; it is much easier for the airline and less transparent for the customers to close a cheap booking class again than removing a lowered price from the market. Lowered prices – except short-term promotions with clear limitations – have a tendency to stay in the market forever, which is by no means what an airline wants to happen.

In fact, the algorithmic nature of the function, combined with the lack of transparency for the client, is one of the key success factors of revenue management. A smart airline with a strong position in a certain market is potentially able to make a non-scarce capacity look scarce to its clients, achieving on average higher fares for the flights than might be expected with full transparency on the client's side. Nevertheless, this type of windfall profit, through non-transparency and an artificial shortage of supply, has become increasingly difficult to achieve in times of meta-search engines, comparing all available offerings in a second.

In any event, the most powerful effect of revenue management does not occur in case of mediocre capacity utilisation but on routes where demand exceeds the available capacity. In this case, revenue management tries to avoid a situation such that low-yield demand displaces the high-yield demand expected later in the booking process. Fostering the allocation of high-yield demand at the expense of low-yield demand – combined with overbooking (see above) – is the most crucial task of revenue management on flights with high demand and utilisation.

9.5.2 The bid price concept

Irrespective of its positive effect on airlines' revenue and profitability, as described above, revenue management is a complex task, with considerable skills needed to perform it in an efficient way. After it had been established at most airlines, it became obvious that its algorithmic foundation required continuous maintenance, if not renewal. For instance, it is in theory beneficial to sustain a large number of booking classes, but in daily practice this means a high effort to keep all relevant data (available inventory, entry price, etc.) up to date. Allocating a single booking request to a seat on a certain flight can become a surprisingly complex process, in which a variety of booking classes and the respective forecasts need to be checked.

Furthermore, many travellers have to book more than one flight to reach their destination, making the allocation decision even more complex; what happens, for example, if a business traveller wants to book a flight from Philadelphia via New York to Paris, and the long-haul flight (JFK – CDG) is wide open in the respective segment, while the required feeder flight (PHI – JFK) is almost fully booked? In the worst case, the client could be rejected for the entire O&D, though his overall profit contribution would be much higher than that of another passenger accepted for the feeder flight, reaching his final destination already in New York. As with the case of network management, such distortions can be reduced by applying correction factors, even though their use is complex and often inconsistent.

Whereas the first aspect (maintenance of booking classes) is a complexity issue, the latter resembles the leg-based versus O&D-based dilemma discussed for network management in Chapter 2. In the late 1990s both challenges were solved at the same time, in the form of the newly introduced *bid price* concept (developed at the aviation faculty of MIT). The idea behind bid pricing is ingeniously simple: instead of checking the access conditions for each and any relevant booking class, revenue managers simply have to check *one single access rate*, called the bid price.

The bid price is composed – on the basis of a mathematical function – out of the access rates for all available booking classes, taking the current forecast for expected demand per segment into account. The application of the bid price model reduces the allocation process to one simple check: if an incoming booking request has a fare higher than the bid price, it will be accepted; if the fare is below the bid price, it will be rejected (or automatically redirected to another flight on that route). This process resembles an auction for a given capacity, which is the reason for the term 'bid price'.

9.5.3 O&D-based revenue management

The even more exciting effect of the introduction of the bid pricing concept was that it automatically facilitated the move from the previous leg-based approach – with the weaknesses described above – to an O&D-based approach. This was the case since bid prices per leg can be easily summed up to a bid price per O&D, removing the risk of losing a high-yield transfer passenger since the feeder or de-feeder flight is blocked by point-to-point passengers with a lower overall value. In the PHI – JFK – CDG example mentioned above, the local passenger (PHI – JFK) is not able to displace the transfer passenger (PHI – JFK – CDG) any more from the last available seat on PHI – JFK if the latter has the higher overall value. The calculation could look as follows (notional figures):

- initial bid prices: US$500 for PHI – JFK, US$2,000 for JFK – CDG;
- initial combined bid price for PHI – CDG: US$2,500;
- passenger A is ready to book a tariff worth US$600 for a flight PHI – JFK;
- passenger B is ready to pay US$2,800 for the itinerary PHI – JFK – CDG (of which, say, US$400 would be pro-rated to the feeder flight in a leg-based revenue management system);
- ex post, passenger B would have had an incremental value of 2,800 – 2,500 = US$300, which is higher than that of passenger A (600 – 500 = US$100). But how to make that information available for an ex ante decision?

The solution is that, if the respective forecasting model is good enough, it will foresee that there will be two passengers 'competing' for the last available seat on PHI – JFK, with passenger B having a higher incremental value than passenger A (US$300 versus US$100). As soon as the revenue management receives this forecast, it would raise the bid prices for the mentioned segments as follows:

- BID price PHI – JFK = US$800 (i.e., passenger A gets rejected with his 'offer' of US$600);
- bid price PHI – CDG = US$2,800 (i.e., passenger B gets accepted for his entire itinerary);
- bid price JFK – CDG (as the difference of the previous two) = US$2,000.

The bid price for the long-haul segment JFK – CDG segment remains stable in this example, which is realistic if that flight has enough capacity left at this point in time. If this flight were short of capacity as well, however, this could change; if there were only one seat left on the flight JFK – CDG as well, and a third passenger C competed with passenger B for this last seat, the new calculation could be the following. Passenger C is ready to pay, say, US$2,500 for a ticket on JFK – CDG. He does 'compete' with passenger B on this flight, but not with passenger A on PHI – JFK. With hindsight – given the initial bid prices – it would now mean more incremental value for the airline to accept passenger A on PHI – JFK and passenger C on JFK – CDG, and reject passenger B:

- passenger A adds US$100 of incremental value on PHI – JFK (600 – 500 = US$100);
- passenger C adds US$500 of incremental value on JFK – PHI (2,500 – 2,000 = US$500);
- passenger B adds US$300 of incremental value on PHI – CDG (2,800 – 2,500 = US$300).

In this case, accepting passengers A and C would lead to an incremental value of US$600, so that passenger B (with a value of only US$300) should be rejected. If the forecasting model predicted this situation correctly, revenue management would adjust the bid prices as follows:

- bid price PHI – JFK = US$600 (i.e., passenger A gets accepted);
- bid price JFK – CDG = US$2,500 (i.e., passenger C gets accepted);
- bid price PHI – CDG = US$3,100 (i.e., passenger B gets rejected with US$2,800).

Obviously, this sophisticated method is fully dependent on the accuracy of the forecasting model. Furthermore, it must be kept in mind that passengers can 'compete' for a seat only if they look for tickets in the same cabin compartment with the same seat configuration. For example, a business-class passenger cannot displace another passenger from the last economy-class seat, since this cannot be transformed into a business-class seat (the respective passenger can of course book a more expensive economy-class ticket, however, and still displace the other passenger).

There are various mathematical methods to predict bid prices. One of the most popular is the 'expected marginal seat revenue' (EMSR) concept suggested by the MIT aviation faculty.[2] There are various other models working efficiently as well, however, dependent on the detailed setting at a certain airline. The example shown above is, of course, a very simple one, with only three passengers competing for three seats on three flights, belonging to three O&Ds. For hundreds of O&Ds, the calculation of bid prices of course turns into a much more demanding task.

This calculation is not really complex by mathematical standards, though, since it can always be reduced to a linear equation system, solvable with standard tools such as a Gauss solver (see paragraph 'Integrated network and crew planning' in section 9.3). The real challenge is to forecast bid prices on a sufficient level of granularity (per single flight) with the sufficient accuracy. A poor-quality bid price is a bad steering instrument and will lead to a costly misallocation of emerging demand to available capacity.

9.5.4 Integrated network and revenue management

The most important integration between network and revenue management has already been achieved: the switch from a leg-based to an O&D-based paradigm for both functions. The times when network management were already capable of digesting the O&D perspective, while revenue management was still forced to take decisions based on a variety of rules of thumb and correction factors, are over since the beginning of the new millennium. A network is most effective if the assumptions on performance indicators taken during the planning and scheduling of the network (such as seat load factors or fare mixes) are realised – if not exceeded – in revenue management. Otherwise, a network looking great in theory may easily become unprofitable in the real world.

The consistent methodology (i.e., the O&D-based perspective) increases the likelihood that revenue management will be able to realise the concept that network management had in mind when designing the schedule. Of course, as conceded above, a weak market or fierce competition will often compromise the work of both network and revenue management; an inconsistent methodology, on the other hand, will *always* do harm to the profitability of a network.

The next step would be a full integration of network and revenue management, which could be achieved, at least in theory, since both functions deal with the match between demand and capacity available in a given network. The more important question, however, is not whether this *could* be done but whether it *should* be done: would that step add any value? The answer is by no means trivial. As long as network design and scheduling follow the current practice of cutting the complex allocation and optimisation task into manageable slices (by resource: see Chapter 4, section 4.3; and along the timeline: see section 4.4), as described in Part I of this book, the integration of network and revenue management is no realistic option. What exactly should be integrated with revenue management if strategic scenario planning and mid-term scheduling are still performed separately, or aircraft rotations are planned – at least partly – independent of crew rotations? On the contrary, in a paradigm with stepwise optimisation, the final step performed by revenue management is by no means a burden but adds value.

Furthermore, revenue management can be perceived as a final sanity check and corrective for network management. As stated earlier, network management optimises – planning-wise – the allocation of available capacity to expected demand, while revenue management assigns actual demand to a given capacity. This *double allocation* or *dual matching mechanism* entails valuable opportunities to check assumptions and correct them if appropriate, taking the most recent changes in demand or competition into account. In this regard, network and revenue management form an iterative loop that enables continuous learning and improvement.

Conversely, a full integration of network and revenue management is thinkable and could generate synergies – since both functions, basically, deal with the same allocation task – but the above-mentioned *dual matching mechanism* and iterative loop would get lost. Furthermore, taking a closer look at revenue management reveals that this function deals with a variety of more operational tasks that are not covered at all by network management; for example, revenue management calculates overbooking rates and booking priorities for flights with very high seat load factors. If for instance, the remaining capacity on a popular flight becomes a scarce good, passengers with loyalty status or high lifetime value could get preferred access to the last seats.

9.5.5 Outlook: new trends in pricing and revenue management

Revenue management currently stands at the beginning of a fundamental redesign, reflecting the change of the industrial framework. Historically, the allocation task was *one-dimensional*: one seat for one passenger. Maximising the seat revenue was an important lever for maximising the revenue of the entire flight. Low-cost carriers established an unbundling trend at the beginning of the new millennium, though, forcing hub and spoke carriers to unravel their service offerings as well. Ever since then most airlines have offered at least one basic tariff without any of the 'fringes' (additional services) formerly included in their comprehensive packages. Additional services, such as lounge access, excess baggage or meals, are still available in some of these basic tariffs, but they need to be booked and paid for on top of the basic ticket, establishing new kinds of revenue streams called '*ancillaries*'.

Reflecting this new practice, revenue management and pricing have to solve a more complex equation: the total fare of each passenger can contain much more than just the basic ticket price, provided that he buys ancillary services on top of his seat. This turns pricing and revenue management into a *multidimensional* task, since every single service and every single revenue element needs to be optimised. As long as the revenue for a basic seat is significantly higher than that for every other service element, the allocation of short-term demand to capacity can still be decided on the basis of the basic fare.

If clients with a cheap tariff allow themselves nice fringes, however, they may reach a higher total value for the airline than a passenger with a more expensive fare but no ancillary services. Consequently, revenue management and pricing managers need to develop their models towards a multidimensional approach whereby every single service element is explicitly managed price-wise – provided that it includes a scarce or perishable resource – and the total revenue potential of each passenger is taken into account for the final granting of a seat on the requested O&D.

The other mega-trend in this area is the use of huge amounts of data by high-performance analytical models. This step was performed first by retail chains, changing the price tags for their goods not every week or day but every hour, or even minute – based on complex price/demand models taking a huge amount of environmental data into account. Airlines try to copy this practice and introduce real-time, dynamic pricing models, leveraging the power of multiple online channels. In parallel, the available information about the buying behaviour and preferences of very small customer segments ('micro-segments', 'digital twins') grows exponentially, thanks to the extensive use of social media and online platforms, enabling personalised, or at least semi-personalised, offerings.

Ultimately, these trends will mean the end of the current algorithmic approach in revenue management, based on a limited number of booking classes. Instead, a next-generation hybrid of dynamic pricing and multidimensional revenue management will probably be able to produce a tailored price for every service item, for a single passenger and at any point in time. This would then be the ultimate version of a bid price, making the traditional fencing and nesting concept of revenue management obsolete.

As in the case of the iterative loop discussed above, the vision of dynamic pricing and multidimensional revenue management makes integration between network and revenue management rather unlikely, since they leverage effects that network management does not deal with.

Notes

1 See https://aviationglossary.com/heavy-maintenance.

2 Belobaba, P. P., 'Optimal vs. heuristic methods for nested seat allocation', presentation to AGIFORS reservations control study group meeting, Brussels, 3 May 1992.

Part III

Trends in network management

As stated in the first part of this book (Chapter 2), the last genuinely disruptive innovation was the move from a leg-based to an O&D-based perspective on transportation networks some 25 years ago. 'Leg-based' in this context means that every single flight segment ('leg') that a client may use on his itinerary is separately viewed and optimised by the respective airline, while 'O&D'-based network management takes the entire itinerary of a client (from origin to destination) into consideration.

Going forward, new trends and technologies may push network management to the next S-curve in terms of performance. Thus, this part provides an outlook on potential game changers, such as new demand forecasting models, new database technologies ('big data'), machine learning and artificial intelligence algorithms and alternative optimisation models.

10 New demand forecasting models

10.1 Key questions and challenges for demand forecasting

Forecasting is a crucial and mission-critical capability throughout the entire process flow of network management, since airlines need to allocate very expensive assets under enormous uncertainty. This statement is valid enough in a normal market environment, let alone in disruptive scenarios such as the COVID-19 pandemic in 2020.

Forecasting can address a variety of questions vital for an airline, such as the following.

- How many passengers will book a certain flight for a certain fare on a certain day?
- Which additional service features (ancillaries) will they buy?
- How many passengers will show up at short notice ('go-shows')?
- How many passengers with a confirmed booking will not show up on the day of operation ('no-shows')?
- When will a certain flight arrive at its destination airport?
- Which irregularities will occur in the course of operations (e.g., AOG or other unplanned maintenance needs)?

In this chapter, the focus lies on the first question, since demand prospects per market, O&D, flight, etc. are key for the design and scheduling of every network. The other questions are similarly important, but more relevant for revenue management and operations control. The first question can be split up into two sub-questions.

- Overall demand: how many passengers will in total use air transportation on a certain O&D/route?
- Specific demand: which share of the overall demand will choose a particular airline offering flights on the respective O&D/route?

The second sub-question has already been discussed in Chapter 5 (sections 5.2 and 5.6), so this chapter primarily deals with the first sub-question. Like network management itself, demand forecasting can be subdivided by time horizon into at least three different steps:

(1) *long-term forecasts*: rough demand for air transportation on markets or O&D bundles; relevant for the design of strategic network scenarios;
(2) *mid-term forecasts*: more detailed demand for single O&Ds, considering specific buying preferences in terms of the timing and routing of flights, service level, alliance, etc.; needed for flight scheduling; and
(3) *short-term forecasts*: the ultimate seat load factor and fare mix for a single flight already open for booking; required for short-term network adjustments and revenue management.

It is an experience-based lesson that the second step is even more difficult to perform than the first and the third ones. Long-term demand forecasts rely primarily on macroeconomic data and trends. On a cumulated level, air traffic is strongly correlated with economic development. If companies from country A do business with companies from country B – and the countries are not too close to each other – there will be demand for airborne business travel between those countries, reflecting the size of the economic exchange. Leisure-related traffic, such as tourism or city travel, is driven by the attractiveness of a certain destination. Finally, business and leisure-related air traffic alike are correlated with the wealth level (i.e., the propensity to pay) in the country of origin. As already described in Chapter 3, various studies underline the positive cycle initiated by air travel, namely the accelerator effect on a country's wealth level: air travel leads to growing national wealth, and growing wealth leads to increased demand for air travel.

Long-term demand forecasts can thus be generated with a relatively high degree of accuracy – at least, as long as a global or regional perspective is maintained (long-term forecasts for single flights would not be reasonable). If both the current exchange between economic spheres and the upcoming trends are known, the future demand for air travel can be extrapolated with common mathematical models. Data can be retrieved from publicly available sources, such as national and international trade statistics, studies of economic research institutes, reports from industry associations and publications of NGOs such as the World Economic Forum.

This approach is valid for normal economic cycles, but also – with some limitations – for times of disruptive change, such as the COVID-19 crisis. Inevitably, however, a long-term forecast is unable to predict the consequences of a disruptive change before the disruptive change itself becomes obvious.

Likewise, short-term demand forecasts are mainly based on extrapolation, in this case on a very granular level, such as a single flight or even booking class. If the booking period for a certain flight is already ongoing, the extrapolation can be based on two pillars: the demand and booking behaviour over time for similar flights in the recent past (e.g., in the previous week), and the booking status for the current flight. In the absence of short-term disruptive events, these two pillars suffice to feed functions such as short-term fleet reassignment and revenue management with reliable forecasts until the final booking status at the day of operations.

Mid-term forecasts are the most challenging instances of demand prediction. In principle, the sources mentioned above for long-term forecasts are still valid and helpful, but the transfer from global demand flows to single routes does not work properly. In other words, the more granular the level observed, the less valid are statistical trends or extrapolations, since specific local influence parameters become increasingly important. Similarly, the short-term extrapolation of booking data does not work in the medium term since actual booking data are not yet available at this point in time.

10.2 Current practice: use of MIDT and booking data

Airlines solve this dilemma by extensively recalibrating macroeconomic data and historic flight data, using their own wealth of experience. New trends such as economic downturns or competitor moves can be taken into account, though disruptive events are almost impossible to cover with this approach. For most hub and spoke airlines, the core input for their forecasting (and their network design in general) is still a database called MIDT (Market Information Data Tapes), available under Sabre's market intelligence service AirVision. MIDT was initially founded as a research entity of the Massachusetts Institute of Technology in Boston, Massachusetts, and it then migrated into a commercial data provision service for aviation clients. The idea behind MIDT data was to retrieve as much ticket data as possible from global distribution systems such as Sabre, Amadeus and Worldspan, combine these O&D data with current macroeconomic trends and transform them into a forecast for future demand on the respective O&Ds.

For many years the MIDT service offered the best demand database and forecast available in the industry, which made it a fairly expensive but popular data source. In the beginning (the early 1990s) another strong argument for MIDT forecast data was that the respective sample was quite comprehensive; most (hub and spoke) airlines filed their inventories with the GDSs, since the global reach of this distribution channel supported their sales efforts in a highly effective way. This last argument has eroded over time, however, because of the rise of the low-cost carriers, which from the very beginning avoided using GDS services because of their relatively high transaction costs.

Therefore, the coverage rate of all GDS databases has declined from more than 80 per cent in the beginning to below 30 per cent nowadays. Nota bene, these figures are valid for scheduled traffic only; charter carriers have hardly ever reported their flights to GDS providers. Even that reduced coverage is still a very good sample, though, so the resulting forecasts are still valuable – provided that correction factors for the non-GDS-based LCC business are built in.

Although demand forecasts based on MIDT data sets are still state of the art, they have three major weaknesses (besides the eroding coverage of air traffic mentioned above).

- Although MIDT data are O&D data, they are based on sold tickets, so they represent an *airport-to-airport view*. For clients living close to a major airport, that may be no major distortion, but there are travellers who dwell far away from any airport, or have a choice between various airports. If someone from, say, Harrisburg, Pennsylvania, has flown from New York to London, then used a rental car to reach his final destination in Birmingham (UK), the MIDT database would count this flight on the O&D New York – London. The true O&D (door-to-door view) in this case, however, was Harrisburg – Birmingham, and, moreover, the traveller could have departed from Washington, DC, or Philadelphia. A forecast relying on the MIDT data alone tends to overestimate the probability that comparable travellers from Harrisburg will use the same itinerary again (airport to airport), even though there are other options from a true O&D's door-to-door perspective.
- The demand for transportation on a certain O&D is larger than the demand for air travel. Whereas flying seems to be the only reasonable choice on long intercontinental trips, *other transportation modes* become relevant as alternatives on short- and mid-haul routes. Furthermore, the preferences between the different modes (air, rail, street, sea) may change in line with budget restrictions and regulatory measures. For instance, in the light of ongoing climate change discussions, national governments in Europe have started initiatives to replace domestic flights by rail trips, fostering the latter at the expense of the former.
- Most hub and spoke airlines establish an inside-out, *hub-centric* perspective in their network planning process, and thus handle the demand forecast and the market share calculation (through the market model) as well. This seems to be reasonable, since the hub is the core production site of their network; a hub-centric view nonetheless blanks out certain passenger flows bypassing this specific hub.

The following example may illustrate the third point above: the forecasting and market model of, say, Air France would try to foresee how many passengers are going to travel on O&Ds between the US East Coast and Southeast Asia (e.g., New York – Singapore), and which share of this traffic flow Air France can attract via its hub in Paris Charles de Gaulle (CDG). In particular, the model will compare a connection on this O&D via CDG with competing offerings via London, Amsterdam, Frankfurt, etc. and evaluate the attractiveness of these services.

If the competing hub is not nearby but, instead, remote (e.g., Dubai), many forecasting and market models become less reliable, because they underestimate the resulting bypass effect (e.g., New York – Dubai – Singapore bypasses

New York – Paris – Singapore). This blind spot effect may even get worse if the O&D looks different for the model, but is in practice exchangeable for the client: e.g., Philadelphia – Dubai – Singapore instead of New York – Paris – Singapore for someone living in Trenton, New Jersey.

In summary, established demand forecasting and allocation models for airlines reflect macroeconomic facts and trends (which they should), but still have considerable weak spots:

- Besides macroeconomic trends, they strongly depend on *historic* and *current booking data.* If this input is unavailable or invalid (e.g., for new routes, or in the case of disruptive events such as the COVID-19 pandemic), the results are almost useless.
- They process *airport-to-airport* data, though real demand occurs on a *door-to-door* basis ('true O&D').
- They focus on demand for air transportation, which may be a *constrained demand* if no appropriate flight exists on a certain O&D. *Unconstrained* demand contains all other transportation modes as well, particularly on short- and mid-haul O&Ds. Ignoring *intermodal competition* may, at the same time, over- or underestimate the real demand for air travel, and is in any event a distortion.
- They basically apply an *inside-out* or *hub-centric* view – i.e., they tend to look at the world from the perspective of a single airline's hub. This view may lead to blind spots with respect to competing hubs, underestimating their bypassing potential.

10.3 Outlook: next-generation demand forecasting models

Airlines try to mitigate these weaknesses by intensive recalibration of their models, as well as by applying various correction factors. The current generation of models leaves room for improvement, however, and has particular shortcomings in the event of non-incremental changes. In contrast, next-generation forecasting models address the mentioned shortcomings:

- less dependence on *historic* and *actual booking data*, having instead a consistent and comprehensive modelling of *needs to travel* and *readiness to travel* based on *macroeconomic relations* between regions and countries worldwide;
- incorporate a *door-to-door perspective* ('true O&D'), instead of an *airport-to-airport* view (as supported by airport statistics and MIDT data);
- take into account *unconstrained demand* (including intermodal competition), instead of *constrained demand* (air traffic only); and
- embody a *comprehensive, outside-in view* on traffic flows and competition between airlines and hubs, instead of having a *hub-centric perspective* (biased by the inside-out view of each airline, and underestimating bypassing potential of competing hubs).

An example of an advanced airline demand forecasting model featuring the improvements mentioned above is 'X-VIA P/C' ('P' for 'passenger demand', 'C' for 'cargo demand') from MKmetric GmbH, a German provider of analytical tools for airlines. Next-generation forecasting models not only achieve higher accuracy in their predictions but are also better prepared for the disruptive environment caused by the COVID-19 pandemic in 2020. Airlines around the globe were forced to massively reduce – if not completely shut down – their capacities within a few weeks in spring 2020. Although this was extremely painful, the upcoming relaunch of flights will be even more challenging, since no airline has ever been in a situation like this. The reactivation of 40, 60 or 80 per cent of the initial capacity, potentially within a few months, is by no means incremental, but the most fundamental relaunch in the history of aviation.[1]

It will be highly seductive for the ailing airline community to reactivate idle capacities as soon as the lockdown is released and demand for air transport seems to be returning. Being too quick in this 'rat race' may ultimately be lethal, however, because ramping up flight operations after the shutdown will unavoidably boost operational costs, which will immediately need to be covered by respective revenues – a task that is intimidating, bearing the enormous uncertainty of the current situation in mind.

Next-generation demand-forecasting models will struggle with this unprecedented disruption as well, but their philosophy makes them more resilient against non-incremental changes than traditional forecasting models. Their lack of dependence on historic booking data, as well as their holistic view on external factors, enables them to include all available information on pandemic dynamics, just like any other macroeconomic driver (or constraint) shaping global air travel demand.

Note

1 Franke, M., 'Airline recovery after the COVID-19 crisis: the real disruption is yet to come', May 2020, M2P, www.m2p.net/airline-recovery-after-covid-19-crisis.

11 New database technologies ('big data')

11.1 The 'big data' vision

Around 2010 a new mega-trend entered the scene – industry-wide, but also in aviation: the introduction of database technologies capable of mastering enormous amounts of data in a very short timeframe. The development of the new technology was based on two fundamental beliefs:

- the results of planning and steering processes improve with more input data, and improve hugely with huge amounts of input data; and
- the ongoing digitalisation of business processes – or even entire businesses – leads to real-time interactions online, which require very rapid response rates by databases and tools.

But what do 'huge amounts of data' and 'very rapid response rates' mean? Since these are technical terms, it would seem reasonable to measure them against the performance parameters of database solutions already existing at that time. Standard SQL (Structured Query Language) databases, such as Microsoft SQL and Oracle 11G, can easily handle data volumes up to 1 terabyte (10^{12} = 1 trillion bytes), achieving response times of a few minutes for complex requests. SQL solutions are more than sufficient for the analysis of operations. Many data warehouse solutions contain so-called data or OLAP (online analytical processing) cubes, offering well-structured access to multidimensional data.

In Figure 11.1, this solution cluster coincides with the bottom left sector of the matrix spanned by data volume and response time. The respective use cases are 'slow and small' – of course, only compared to real 'big data' cases (see below). In this cluster, input and output data can regularly be processed in a sequential way.

The top left sector of the matrix in Figure 11.1 contains use cases with the label 'small, but fast'. For these cases, very short response times (a few seconds or less) are required, which would be too challenging for traditional databases, because this means that output data must be generated while input data are still being digested. The respective cluster of use cases is called 'stream analytics', describing applications for which outputs must be generated almost in real

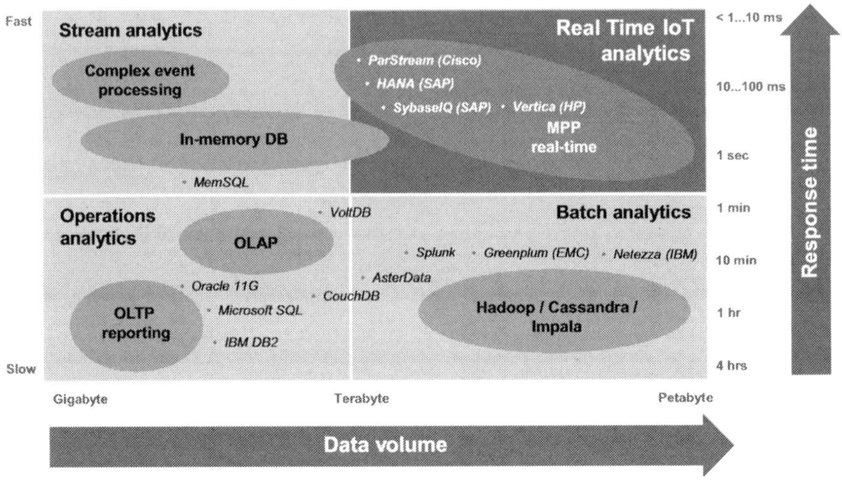

Figure 11.1 Clustering of database technologies for various analytical use cases[a]

Notes: MPP = massively parallel processing; OLTP = online transaction processing.

[a] Bienert, J., 'Company presentation: ParStream GmbH', 2016 [not published].

time (e.g., real-time bidding on commercial websites). To achieve this speed, the respective solutions need to keep all relevant input data in memory (as opposed to traditional databases, in which the query algorithm has enough time to switch back and forth between database and memory). Consequently, the respective technology is referred to as 'in-memory database'.

The bottom right sector indicates the opposite case: 'slow, but large'. The respective use cases work well with the speed of standard database technology but require the handling of much larger data sets – from terabytes up to petabytes (10^{15} = 1 quadrillion bytes). These volumes cannot be processed by standard database solutions any more but require so-called 'batch analytics'. The input data are digested in a sequential way, but in parallel batches, using special protocols and processors capable of parallel computing. A typical use case in this solution cluster could be a pricing model for retail companies, evaluating billions of relevant data (such as external drivers or cross-elasticities) to find the optimum price for millions of product categories at thousands of shops.

A popular representative of this technology is the Hadoop framework developed by the Apache Software Foundation as a freeware for highly scalable database applications, usually run on cluster computers. The respective database is called Apache Cassandra, a column-oriented NoSQL solution. Cassandra is, for example, deployed by Facebook, Apple and Twitter to handle billions of user accounts.

To become a real 'big data' case (top right sector of the matrix), applications must by definition require both data volumes above one terabyte and response times below one second – i.e., they are 'fast and large'. Popular representatives of this cluster are SAP's HANA, Cisco's ParStream and HP's Vertica solution. The outstanding performance can be achieved only because these solutions combine in-memory computing with column-oriented NoSQL parallel processing (massively parallel processing). Outputs can be generated while input data are still flowing in – a prerequisite for extremely complex, real-time applications. Typical use cases for this leading-edge technology come from the sphere of the 'internet of things' (IoT), the fully automated, continuous and real-time interaction between complex technical systems (machine to machine: M2M).

Most consumers have heard about the popular IoT example of private refrigerators being connected via the internet with the service centre of the household's favourite supermarket chain, automatically sending replenishment orders whenever its sensors foresee the shortage of a certain food item stored in the refrigerator. Although the example is practical and instructive, it is unfortunately neither a 'fast' nor a 'large' case, even if millions of refrigerators reported their needs to a single service centre. This task could still be mastered by standard SQL databases.

Obviously, IoT applications may, but do not necessarily have to be, 'big data' cases. More likely to fall into this category are IoT applications in complex production lanes (e.g., operated by car manufacturers) or process plants (operated by chemical companies). If thousands of interdependent production steps with millions of state variables need to be supervised and – if required – adapted every second, the automation of such plants could ultimately require big data technology.

Some experts are even more demanding; they accept the label 'big data' only for applications requiring petabytes of data and response times as fast as milliseconds. This refers to 'extremely fast' and 'extremely large' use cases, in the upper right corner of the top right sector of the matrix in Figure 11.1. The respective applications are extremely advanced and exhibit performance at the absolute rim of what computer technology can do today. A real-life IoT use case realised with ParStream's big data technology can be found in Appendix B.

11.2 Big data applications at airlines

The obvious question now is whether advanced big data technology can create value at airlines, and especially in network management. The simple answer is 'Yes', there are useful applications at airlines (e.g., dynamic pricing: see below), but 'No', not in network management – at least, not in the current setting. As with the case of other technological innovations, proponents tend to hype their favourite topic, furthering the impression that it is a one-size-fits-all silver bullet for each and every business challenge. This is, of course, not the case. As described in the preceding section, the ideal use case for big data solutions requires the processing of more than one terabyte (if not petabyte, in the more rigid definition) with a response time of less than one second.

These requirements do not exist in network management. Network management scenario calculations look complex to businesspeople, and so they are – particularly for major international airlines, with hundreds of aircraft serving hundreds of destinations, with tens of thousands of potential combinations (i.e., O&Ds). This type of complexity is far away from the systemic thresholds defined above, however. The allocation of demand and capacity to routes in such a way is beyond anything that a human brain could ever handle, but it can regularly be handled by traditional linear programming solvers. Even if thousands of travellers are available per O&D every day, and large airlines cover tens of thousands of O&Ds, the data volume to be handled at a time achieves a magnitude of megabytes, or maybe gigabytes, but by no means terabytes, or even petabytes.

Furthermore, network management tasks are by definition not real-time tasks, since planning processes are performed weeks and months before the actual flight event. So, even if – for whatever reason – the amount of data to be digested for the design of network scenarios or schedules were to grow by several orders of magnitude, they could still be managed with NoSQL parallel-processing databases such as Hadoop or Cassandra (bottom right sector in Figure 11.1).

If not network management, which other management function of airlines could benefit from or even require big data technology? The most likely candidate is dynamic pricing. In theory, the accuracy and efficiency of both pricing and revenue management can be increased ad infinitum by providing *more* and *more actual* input data. For instance, macroeconomic trends, the current offerings of competitors and other decision criteria can be transformed into billions of price/demand curves with trillions of cross-elasticities, until, ultimately, every single client receives a fully personalised price at any point in time. A pricing machine responding to respective booking requests in real time, taking trillions of actual price triggers into account, would definitely require big data technology.

The only exception from the judgement that network management will not require big data technology would be the fictitious scenario described in Chapter 9, section 9.5, in which network management, pricing and revenue management are merged into on integrated super-function, steering the allocation of capacity to demand, and demand to capacity, from long-term strategy all the way to very short-term adjustments, including the segmentation of capacity and demand. This is a long-shot scenario, however, that will not materialise before 2025.

12 Machine learning and artificial intelligence

12.1 The pedigree of machine learning solutions

As with big data in the previous chapter, artificial intelligence (AI) has been a very popular topic – if not a hype – over the past couple of years. And, likewise, AI opens up a universe of new opportunities across many industries. But it is definitely unable to do magic. Many people expect that AI can solve optimisation problems better than traditional methods, or solve problems that could not be solved *at all* up to now. This is not what AI is about, though; AI is not primarily an optimisation method for deterministic problems. Before elaborating on which amazing things AI can do instead, it will be helpful to clarify the terminology first.

Figure 12.1 shows the mathematical pedigree of AI. First of all, it needs to be noted that, if people talk about *AI*, they very often mean *machine learning*, of which AI is a subset. Thus, this chapter deals with machine learning, not just with its derivative, AI. Starting from the top of the pedigree, machine learning is one of the three major methodological areas of the mathematical domain *data science*, the other two being *data mining* and *statistics*. The latter two represent the traditional approach to extracting maximum value from available data; data mining deals with smart ways to collect data, and statistics provide tools to identify structures, patterns and trends in data sets. In contrast, machine learning is a relatively new sub-domain of data science, aiming at establishing more dynamic solutions that grow with the task.

As indicated by the name, machine learning tries to mimic the capability of the human brain to find new strategies for solving new problems, based on experience and learning. Learning is a key success factor of every being, because it massively expands the range of potential solutions, adjusting them to a changed environment. Compared to this ideal, a method such as statistics remains static, unable to create solutions beyond the scope carved into its DNA through the initial set of rules. For instance, a moving average will always be a moving average, and, if this method does not lead to a useful insight about a data set, statistics will not automatically provide an alternative method better suited to find the relevant pattern.

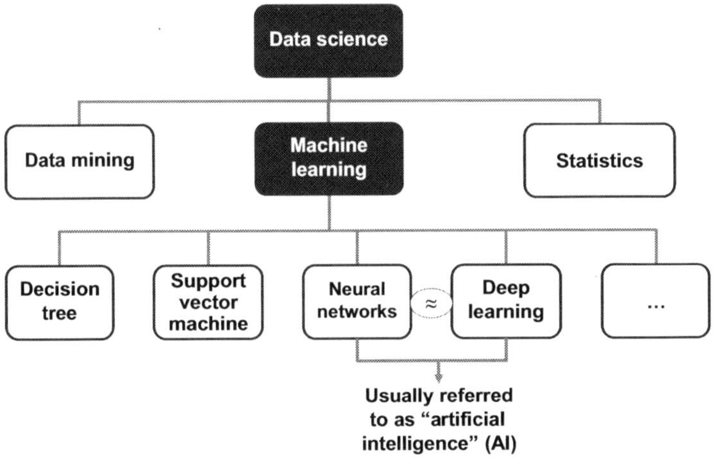

Figure 12.1 Typology of machine learning methodologies[a]

[a] Plutecki, M., 'Company presentation: Valculus GmbH', 2017 [not published].

12.2 Historic roots of machine learning

Machine learning algorithms are designed to go beyond the initial rules and find new ways to tackle a changed problem. Like the human brain, these algorithms need to be trained with suitable inputs, forming an 'experience' base that the algorithm can subsequently leverage for new solutions. The algorithm can thus become smarter over time and solve problems it would have been unable to solve in the beginning. The mathematical theory and toolkit behind machine learning is, by the way, not at all brand new, but 50 to 60 years old; there are various reasons why it took that long to make it a mega-trend, though, and extract value from it in business practice:

- In the 1960s, when the first machine learning theories where created by mathematicians (the term 'data scientist' was not yet invented then), the level of digitalisation of life was close to zero, so no electronic data were available that could have been processed by computers to analyse business problems.
- Likewise, business processes were not at all digitalised, so, even if electronic information had been available, a machine learning algorithm would have had no opportunity to have an influence on real-life business decisions.
- Last, but not least, computers were not omnipresent, but very rare tools in the 1960s, with a tiny fraction of the computational power available today. Thus, complex algorithmic calculations could not be performed quickly enough for machine learning use cases, or be performed at all.

Today the mentioned prerequisites are in place, so that machine learning could become the hype topic that it is nowadays. Machine learning uses a variety of methodologies, such as 'decision tree', 'support vector machine' and 'deep learning', the latter being the most recent and most well-known occurrence. It has particularly become popular under the alternative name of artificial intelligence. Some experts also use the term 'neural networks' as a synonym for AI or deep learning as well, but neural networks are the method behind deep learning rather than an alternative denomination.

In any event, neural networks are the consequent continuation of the initial vision behind machine learning: they copy the flexible way that human brains use to find solutions for new problems based on rules and experience. As in a brain, a mathematical neural network is capable of dynamic growth, creating new connections (nodes and edges) under the impression of new challenges.

Thus, a deep learning or AI algorithm – or even a machine learning algorithm, on the next level above – can, in theory, learn to solve every single task that a human brain is able to handle. In practice, this capability is contingent on two prerequisites: that the way a human brain tackles the task can be described in a structured way, and that enough appropriate training data are available.

Since machine learning and AI are hot issues, many scientists and industry practitioners have suggested typologies for tasks that can be performed by machine learning. The following five task clusters have been defined in the context of a digitalisation study for airlines, prepared at the private university WHU in Vallendar/Koblenz, Germany:[1]

(1) analyse *structured data*, find *patterns* (beyond obvious patterns that statistics could identify);
(2) *predict* future events or future (human) behaviour based on identified historic patterns;
(3) recognise *unstructured data*, such as text, images, speech, faces, sentiments and sounds;
(4) *react* to recognised *unstructured data*, such as messages or events, in a structured way (e.g., provide answers to questions); *interact* with human beings in writing or artificial speech; and
(5) use all the capabilities above to move things (robotics).

12.3 Machine learning use cases

To fully understand the benefits of the new tools, a brief comparison between traditional approaches and machine learning algorithms for the first three tasks listed above may be helpful:

- pattern detection
- prediction
- text recognition

Pattern detection in structured data is definitely a task that statistics models can tackle with the tool of correlation analysis. If an airline wants to better understand why certain clients book certain flights or buy certain ancillary services, they can screen all available historic transaction data, compare them with basic client data and try to find correlations between them. The weakness of this approach is the limitation of human brains; the analysis requires a human operator formulating hypotheses for the expected correlation, say that clients with a certain age, home address, salary level or transaction history are more likely to book a specific flight or service than others.

The limitations are that human experts will formulate only hypotheses that they find logical and they will restrict the potential triggers to a few (studies say that human brains cannot handle more than seven different aspects at a time). Likewise, the segmentation of clients will usually lead to intuitive clusters that every expert would have expected anyway; this can easily result in a self-fulfilling prophecy type of bias.

In contrast, a machine learning algorithm is able to analyse available data totally independently of these typical human limitations or biases. Once an initial set of correlation hypotheses is provided by human experts for training reasons, the self-learning algorithm can check thousands of triggers with millions of correlations at very high speed, creating new hypotheses through the 'mutation' of existing ones whenever appropriate. Consequently, the algorithm may define client segments that no human brain would ever think of and find patterns that no statistics tool would ever be able to detect.

On the other hand, this strength is at the same time a potential weakness of the machine learning approach: the results of the analysis – with respect to client segmentation or correlations – may be so counterintuitive that human experts are unable to understand the underlying logic. If, for example, a machine learning algorithm declares that female clients aged between 29 and 32, preferring beer to wine and speaking at least three languages have the highest likelihood to book an upgrade on a trans-Atlantic flight, while divorced male clients aged between 48 and 53 with at least two kids have the lowest likelihood, this is a stunning result – but how should salespeople deal with the insight? And is this segmentation valid for any other correlation analysis? In any event, one of the strengths of machine learning algorithms is that they can be trained to detect correlations between hundreds of parameters when every statistical method would inevitably fail.

The *prediction* of future events or behaviours usually leverages patterns detected in historic data (as discussed in the previous paragraph), adding expectations about trends in external or internal triggers. In this respect, forecasting is always a kind of extrapolation of the past. If historic data are unavailable (e.g., on a new route), or disruptive events spoil the perceivable trends, the results of forecasting methods will be poor.

This is, basically, also the case for prediction models based on machine learning; on the other hand, the latter algorithms can take into account a much larger set of triggers and (expected) correlations (as with the case of pattern finding discussed above), so that the respective predictions can achieve a higher accuracy both under regular or disruptive conditions. The predictions will be less an extrapolation of historic data and more a highly sophisticated, multi-variable trend analysis – as with the case of *pattern finding*, with a huge number of triggers and correlations that no human brain could ever handle.

As an example, a start-up company has recently recorded millions of sound samples from aircraft jet engines, testing the correlation between certain noise types, frequency patterns and subsequent failures. After a comprehensive training phase, the machine learning model learned to predict the likelihood of an engine failure at every point in time based on a short sound check of the engine – with amazing accuracy. No traditional, static forecasting model could handle the complexity of this problem and come to a reliable prediction of engine failures based purely on engine noise.

Text recognition – i.e., the understanding of language in written form – is a capability that human beings learn over many years, the proficiency level increasing from single words via simple sentences, more complex grammatical constructions and sophisticated semantic values to unwritten connotations, such as irony or sarcasm. Traditional algorithms for text recognition rely on a number of key words customised for a specific topic, hoping that this limited vocabulary suffices to roughly grasp the meaning of a text. This approach may work for simple, standard messages, but is definitely overstrained with more complex structures, let alone with implicit facets of a language such as irony.

A typical use case of text recognition for an airline could be the automated screening and answering of incoming client complaints. Such messages are numerous, annoying to process and time-consuming, so an ideal use case for an automation solution. Nevertheless, clients complaining about a perceived underperformance or impertinence are usually emotional, use rude or sarcastic language and often address more than one issue. Traditional, key word-based (semantic) algorithms are simply unable to solve this task, and, even if they could, they would not be scalable for larger numbers of messages or topics.

The right choice are deep learning algorithms, because they can learn language like a human being: as a complex code conveying semantic meanings, not as a linear sequence of words. The respective domain of deep learning is called 'natural language processing' (NLP). As stated above, neural networks, as core elements of deep learning models, mimic the way that human brains create new connections, and thus expand their capabilities. After extensive learning, such an algorithm can grasp the meaning of complex messages, even under aggravated conditions such as typos or irony, and achieve scalability for almost every case number.

The caveat of this method is that the neural network has to be trained not only for the language itself but also for the specific communication situation; otherwise, it will still fail. An algorithm that recognises complaints for an airline with high accuracy will need to be retrained for complaints in other industries, since the respective communication patterns and topics are different.

Besides the pedigree illustrated in Figure 12.1, there are other options to cluster machine learning algorithms. The following paradigms differ in the way that the learning effect is triggered:

- *supervised learning*: this is the base case, as explained so far in this chapter. The machine learning algorithm is educated with labelled training data, consisting of clearly defined input data and a desired output value. After a sufficient number of training data, the supervised learning algorithm will be able to understand the functional connection between input and output ('mapping'), and transfer that to previously unseen input data.
- *unsupervised learning*: this tackles the same task as supervised learning with similar algorithms, but with much less (semi-supervised), or even no, labelled training data. Respective solutions have to perform a so-called self-organisation, calculating the probability densities of input data. The algorithms learn as well, but, in the absence of human-labelled data, more slowly than supervised algorithms (e.g., through cluster analysis).
- *reinforced learning*: unlike the two previous paradigms of machine learning, reinforced learning does not need labelled pairs of input/output data, nor an exact mathematical function for the correlation between input and output. Instead, the respective algorithms continuously evaluate the right balance between existing knowledge about the problem and unseen occurrences using dynamic programming techniques. In this regard, reinforcement learning is a non-deterministic approach. The environment of the problem is usually described as a Markov decision process (MDP).

It is well known that machine learning models *are by no means* smarter than traditional, deterministic models in solving the complex equation system of target function and constraints, but they are much better prepared to understand the impact of dozens, if not hundreds, of parameters on the efficiency of a solution. In contrast to the traditional models, for which the optimum mix of parameters is an *explicit* result of the relevant equations, the machine learning model identifies efficient solutions based on an *implicit* understanding of the drivers.

12.4 Machine learning applications in network management

As stated earlier in this chapter, machine learning has not been developed to replace traditional optimisation models – such as linear programming as the preferred solution for the demand and capacity allocation problems occurring in network management – but to automate complex and dynamic decision processes that previously required human intelligence. Once the machine learning algorithm has understood – through extensive training or other mechanisms – how a human brain would define the functional connection between input and output data, it is able to transfer this insight to previously unknown input data, embedded in a continuous self-learning process.

Although solving the allocation problem itself is not a typical application for machine learning, there are nonetheless adjacent tasks of network management to which machine learning algorithms could add value:

- prediction
- scenario definition and calculation

As claimed earlier in this chapter, *prediction* is one of the key applications of machine learning algorithms. This does not primarily include demand forecasting, however – a task that can be tackled with traditional econometric models (see Chapter 10). Instead, self-learning solutions and neural networks are well prepared for a more detailed prediction of clients' behaviour. Respective models are able to detect even complex patterns in historic transaction data and foresee potential future client preferences.

In other words, the question solved with a machine learning model would not be 'How many clients want to travel on a certain route at a certain time?' (i.e., the macroeconomic perspective) but which buying preferences (e.g., for booking classes or ancillary services) certain client segments could have on a specific flight (the microeconomic perspective). In this regard, the machine learning solution would not be applied in the core of network management but, rather, at the interface between network management, pricing and revenue management.

Most of the optimisation methods for transportation problems described in Chapters 5 and 6 require the *definition of scenarios* as a prerequisite, usually by human experts who know the detailed business environment for the optimisation problem. This task can become a bottleneck of the process, since it is often time-consuming and demanding to formulate appropriate scenarios or hypotheses. A neural network could help: if the model can be trained with information on how human experts build and evaluate viable scenarios, it may learn how to automate this task – provided that the functional connection between inputs (economic goals, industry background, available building blocks of a scenario) and output (viable scenario) can be clearly defined. Building a scenario can be a key step for strategic network planning, but also for short-term recovery measures (i.e., for disruption management).

Note

1 Cascio, E., Ringbeck, J., and Franke, M., 'The future role of AI/business analytics in marketing and sales: an empirical investigation in the European airline industry', master's thesis, WHU, Vallendar/Koblenz, 2019 [confidential].

13 Alternative optimisation models

Up to this point, it has been presumed that the demand and capacity allocation problem posed to network management can be tackled with traditional, deterministic, holistic optimisation models such as linear programming. The proof of this assumption is adduced every day in the network management departments of major airlines around the globe. The solutions generated with these models come at a price, however: finding the perfect solution for a given allocation problem can be very laborious and time-consuming.

Thus, it is reasonable to question the value-for-money ratio of solvers searching the global optimum for a given allocation problem. In the case of iterative solutions, this ratio can easily be checked by calculating the incremental value and the incremental effort for the next iteration step. If, for example, an algorithm converges towards the perfect solution with a value of X, requiring 1,000 iteration steps, and 90 per cent of this value can be achieved after only 100 iteration steps, it may be reasonable to define the 90 per cent as the accuracy threshold and stop all subsequent iterations after 100 steps – provided that a 90 per cent accurate result represents sufficiently high quality for the given use case.

But how to apply this pragmatic approach to more comprehensive, highly complex optimisation models? There are various mathematical theories dealing with this question; the following three examples are meant to illustrate the thinking behind such concepts:

(1) network slicing (partial solutions)
(2) non-deterministic algorithms
(3) quantum computing

Network slicing is, as the name indicates, a mathematical concept developed for network structures. It was initially designed to reduce the complexity of mobile phone infrastructures. In times of cloud computing, smartphone users simply assume that they have seamless access to every piece of content irrespective of their current location. To meet this expectation, mobile carriers need to realise the so-called information-centric networking (ICN) paradigm, which ensures high performance, but causes very complex networks with high investment needs (capital expenditure: CapEx) and operational costs (operational expenditure: OpEx).

Network slicing aims at achieving almost the same performance with less infrastructure, distributing content across the entire network and reducing the burden of central network instances by using decentralised caches in cell base stations, and even on user devices. The latter method is referred to as fog radio access network (F-RAN), which is basically a fog computing application[1] (see the fog computing example in Appendix B).

The question now is the degree to which the mobile phone network case can be transferred to airline networks. First of all, 'content' in the case of air travel is more physical than for mobile phone users, featuring, for instance, seats, meals and baggage transport, so the 'content' cannot be stored, and the fog computing paradigm does not apply. The slicing of networks is still relevant, however, since it could reduce the complexity, and thus planning costs, compared to a holistic optimisation approach. Hub and spoke carriers already today slice their networks into up to three layers:

- long-haul services (using wide-body aircraft);
- short- and mid-haul services (using narrow-body aircraft); and
- regional services (using small regional jets).

The separation is easy in this case, since the respective sub-fleets are different, so that the touchpoints – both physically and in terms of planning – can be reduced to connections in hubs. A really beneficial 'slicing' approach would have to go beyond this obvious distinction, however. Every airline would have to check whether 'network slicing' could provide a system of sub-networks that can be optimised with limited effort using traditional methods, and lead to a sufficient level of performance and efficiency for the overall network. Or, in other words: will a system of local optima be close to the global optimum, or significantly inferior?

For experienced network managers, this compromise is somewhat awkward, since they have been raised with the paradigm that moving from a single leg perspective to the entire network perspective is the only way to correctly allocate revenues and costs, and thus to find the one and only optimum network scenario. Everything in between is no man's land for them. The slicing approach can still be valuable, though, if the performance and network result stays high but the planning effort goes down. This concept would, by the way, fit particularly well with the potential change towards a less determined, more flexible network planning paradigm, assigning resources as late as possible in the process, as outlined in Chapter 4, subsection 4.4.14.

Non-deterministic optimisation algorithms are, primarily, a theoretical construct used in information sciences. They have already been described by early IT pioneers, such as British mathematician Alan Turing. The term 'non-deterministic' in this context means that one input value (or vector) can lead to more than one output factor. The correlation is either purely random or labelled with a certain probability (similar to *stochastic optimisation*). The probability may reflect knowledge about the changing environment, about the target function or about the potential outcome of the optimisation problem. This concept distantly resembles the quantum-mechanical atomic model: although the energy level of a single atom (and, thus, of its electrons) may be well known at a certain point in time, the exact position of the electrons on the outer shells of the atom is by definition unknown (a fact postulated by German physicist Werner Heisenberg, and thus known as the 'Heisenberg uncertainty principle'). This leads to the counterintuitive conclusion that a single electron circling an atomic nucleus can be at different spots on his trajectory at the same time – the only restriction being that the probabilities for the different locations sum to 100 per cent.

As mentioned above, non-deterministic algorithms are primarily used by information scientists and IT experts to estimate the upper boundary for the effort related to deterministic optimisation models. It is often easier to find a non-deterministic algorithm for a complex problem than a deterministic one. Unfortunately, non-deterministic algorithms cannot be realised on traditional computers, so the insight that a non-deterministic algorithm does – in theory – exist for a certain optimisation problem does not directly help solve the problem in practice. It *does*, however, help estimate the effort needed to solve a previously unknown problem.

For example, if *no* non-deterministic algorithm can be found suited to (in theory) solve a known problem A, and the effort to solve this problem is X, this can be used as the upper boundary for the effort to solve a similar, previously unknown problem B – provided that a non-deterministic algorithm can be found for problem B. This correlation is proved, since problems with at least one appropriate, non-deterministic algorithm are always of a simpler kind than algorithms without this phenomenon.

It remains to be seen whether future computers are prepared to execute non-deterministic algorithms; if so, these models could be used also to solve network management optimisation problems at higher speed with lower effort. Till then, this class of algorithms can at least be used to estimate the complexity of an unknown optimisation problem, as sketched above.

Quantum computing is, first of all, a new hardware concept for computers, but, once machines of this type work in practice, this concept will also revolutionise software applications such as optimisation models. Like traditional computers, quantum computers use a binary code of two clearly distinguishable states as their basic information entity. Whereas traditional computers use *electrodynamic states* to define the binary code (referred to as a bit: 0 = current off, 1 = current on), quantum computers try to use *quantum-mechanical states*, the so-called qubits.

The practical problem of this concept is that quantum-mechanical states tend to be much more unstable than electrodynamical states, which makes it extremely difficult to store or process digital information with them. The current record (2020) for quantum machines is a capacity of 70 qubits under laboratory conditions, which is, of course, tiny compared to traditional supercomputers with parallel processing capabilities prepared to handle terabytes up to petabytes (10^{12} to 10^{15} bytes) of data.

The intriguing perspective about developing quantum computers is that they will be able to process information based on quantum-mechanical principles such as superposition or quantum entanglement – effects that will open up a completely new world of applications. For our concrete application, the optimisation of a logistic (allocation) problem, the following approach would become thinkable with a quantum computer.

Let us assume we have a very simple problem with a target function (reflecting, for instance, the overall profit on a certain route), and only two variables (e.g., the number of aircraft and the size of the aircraft). We look for the optimum constellation – i.e., the solution with the highest profit. This problem could be illustrated as a three-dimensional graph, with the two variables forming a horizontal plane, and the resulting profit being depicted on the vertical axis. The resulting graph (i.e., the solution space) will look like data mountains, with 'valleys' (poor solutions) and 'peaks' (good solutions). The optimum solution with the maximum profit is the highest peak of this solution space.

All traditional solvers on traditional computers basically use the same 'stupid' approach to find this global maximum: they walk up and down the 'mountains' until they have found the highest peak. The intelligence of the different methods (and, thus, the resulting efficiency and speed) lies in the smartness of their mapping and pathfinding approach – i.e., how quickly the algorithm realises that it is walking in a good or bad direction. (This is, by the way, very similar to the approach of LP algorithms described in Chapter 5 and Appendix A, moving around a solution polygon to find the corner with the highest target function value: see Figure A.1).

A solver on a quantum computer would be much quicker and more efficient, because it could leverage quantum-mechanical effects. In the concrete example, the solver would not necessarily have to walk up and down the 'mountains' of the solution space, but – in very simplified words – drill tunnels through them – which is, of course, considerably quicker. This approach resembles the thought construct of the so-called 'wormholes' in astrophysics: from a relativistic perspective, our universe is not straight but contains curved areas – nota bene in four dimensions (including time as the fourth dimension). Whereas a spaceship with a traditional rocket engine has to follow the curved trajectory and loses time on this 'detour', a spaceship with a relativistic propulsion system could, in theory, create its own 'wormholes', and use them as short cuts through the curved area to achieve speeds even higher than the speed of light – at least, relative to the curved coordinate system.

For the two-dimensional example discussed above, the time saved through the quantum-mechanical short cut approach may appear limited, but it needs to be kept in mind that most optimisation problems feature many more than two variables, leading to highly complex multidimensional solution spaces (n+1-dimensional for n variables).

As stated above, quantum computers are not yet ready for practical applications, but IT scientist and large IT companies are already diligently working on algorithms and solutions that *could* work on quantum computers, probing the ground for future generations of systems. This branch of IT research is referred to as 'quantum-inspired computing' (QIC). As with the case of non-deterministic algorithms described earlier in this chapter, quantum-inspired computing can generate practical value already today, even before the respective hardware is available. For example, QIC may help find upper or lower boundaries for the speed of traditional solvers, or measure how close existing methods get to the theoretical performance level of quantum algorithms.

The three examples of new mathematical theories raised in this chapter are no more than a hint at what may be available for solving complex network management problems in the future. For most optimisation problems, the traditional solution methods discussed in this book are still appropriate and sufficient. On the other hand, time is money, and mathematical research will proceed in any event, so it may be helpful to have a basic understanding of what advanced theories and methodologies could potentially do for network management practitioners in the future.

Note

1 Jin, H., Lu, H., and Zhao, C., 'Content-oriented network slicing optimization based on cache-enabled hybrid radio access network', *Journal on Wireless Communications and Networking*, 2, 2018, https://doi.org/10.1186/s13638-017-0995-z.

Overall conclusion

This book has provided a comprehensive overview on state-of-the-art network management practices for airline professionals. Thus, a practical view on the matter has prevailed, though some academic perspectives were added to create a better balance and provide a more rounded picture.

The significance and value of network management is undisputed, particularly for hub and spoke airlines. Network operations generate the core product of airlines and determine most of the available assets and resources, so it is intuitive that airline networks account for the vast majority of airline revenues and costs. To run an airline profitably, it is indispensable to design an attractive network and operate it efficiently.

Designing and operating networks is a highly complex task, unless only a handful of aircraft and destinations are involved. This is particularly true for hub and spoke airlines, but it applies even to low-cost carriers, which try to keep their networks simple. The complexity stems from the large amount of interdependent resources, opening up choices in abundance to assign resources to network scenarios, and bring passengers from their origin to their destination. In addition, aviation is still a highly regulated industry, with countless restrictions and legal requirements. Last, but not least, competition is fierce in this industry, since severe overcapacities are a fundamental problem in most parts of the world, while airport and airspace capacity are rather scarce.

Although both airlines and specialised IT solutions providers have invested billions of dollars into data analysis and operations research capabilities over the past decades, there is still no fully integrated network planning capable of translating all framework parameters and strategic goals into a single optimum network solution in one step. Even leading airlines have established network planning practices relying on stepwise, iterative optimisation loops for network schedules, aircraft rotations, crew rotations and hub bank structures. This book has shown that this phenomenon is not primarily attributable to a lack of OR capabilities or computer speed, but to three pragmatic factors.

- The business and the regulatory environment may be very different even between similar airlines in the same geography; a perfect network solution for one airline can definitely be suboptimal for another airline.

- The market environment and competition may change every day, so sticking to an – ex ante – perfect schedule can, ultimately, lead to a disaster if the environment has changed in the meantime.
- Calculating the perfect solution for a highly complex optimisation problem can be very time-consuming and expensive; cutting the problem into manageable slices may significantly reduce the effort and increase the agility and flexibility of the planning process.

Consequently, there is no single 'optimum' network design but, rather, more efficient and less efficient concepts for a given set of external and internal framework parameters. For instance, with a certain existing fleet, a night curfew in the airline's hub and specific labour agreements, one carrier may end up with a significantly different network design from another carrier of similar size but with a different fleet, no curfew and different labour agreements. Yet both designs may be perfectly reasonable and suited for their respective economic environments. This phenomenon seems to be valid not only for the past and present but also for the foreseeable future, since the underlying industry mechanics remain unchanged.

In essence, designing, scheduling and running aviation networks in an efficient way will always require three things: extensive market and traffic flow data; world-class analytical support (e.g., state-of-the-art OR models); and sound experience with the former two. Network schedules will be the more robust and efficient, the more the specific environment of the airline and the interdependences between the required resources are taken into account.

This book has also provided some evidence for the thesis that not every trend being discussed as a 'hot issue' will necessarily make it into tomorrow's network management practice. In some cases there is no match between the needs of the function and the characteristics of the new technology (e.g., big data), whereas in other cases the practical application of a new technology is still quite far in the future (e.g., non-deterministic algorithms, quantum computing).

The same is true for the further integration of network management with its neighbouring functions, such as crew planning and revenue management; in theory, the respective tasks and calculations could definitely be merged, but, if the resulting value does not cover the significant investments for doing so, it will probably not happen. Furthermore, integrating functions with very different time horizons – from a long-term strategic perspective, via mid-term scheduling, to short-term adjustments – would probably compromise the inherent flexibility of the current practice – i.e., its ability to react to changes of the external or internal setting throughout the planning cycle.

For example, key success factors of the current network management paradigm are the stepwise approach (see above) and the *dual matching mechanism*: first – in the long and medium term – the available capacity is adjusted to the expected demand (e.g., by growing or shrinking the fleet); then – in the short term – the actual demand is allocated to the given capacity. This mechanism is instrumental to ensuring good utilisation of the expensive asset 'aircraft', and will probably remain a key pillar of network management.

Despite the mentioned reservations, there are, for sure, significant changes ahead. *First*, the business models of hub and spoke carriers and low-cost carriers will converge further, resulting in new and more hybrid network design patterns. For example, hub and spoke carriers and low-cost carriers may start collaborations, such that the low-cost carrier feeds the long-haul network of hub and spoke carriers at highly competitive costs.

Second, rapidly increasing computing speed and analytical capabilities may bring new and more efficient network design algorithms. The quality and reliability of demand forecasts will increase further thanks to the capabilities of a new generation of holistic forecasting models, and the interpretation of complex network scenarios may be supported by advanced machine learning algorithms.

And, *third*, the ongoing digitalisation of all industries will change the rules of the game for airline network management too: social media, as well as the abundance of online channels, will create new interfaces and touchpoints with customers, leading to an exponential growth of insights into clients' travel needs and preferences at any point in time. The improved transparency for what clients need and want will have an impact on how and when network scenarios and schedules are designed, as it will probably push the focus of network management from long-term scenarios to more short-term, reactive types of schedule adjustments.

Appendix A: Additional information on network optimisation models

Linear programming for network capacity and resource allocation

Network planning problems at transportation companies are capacity and resource allocation problems. Demand (for seats on a certain O&D) needs to be matched with a suitable supply of capacity on routes suited to cover the O&D (point to point/non-stop or with one or more transfers). Since a network consists of a well-defined combination of edges and nodes, a transportation company such as an airline has a choice – at least in theory – to take very different itineraries through its network, and change the allocation of demand to capacity at every node for the subsequent edge(s).

The complexity of the network, and thus the number of different options to move a passenger or shipment from one point (the origin) to another point (the destination) in the network, grows parabolically with the number of nodes, since n nodes can create a maximum of $\frac{1}{2} n \cdot (n-1)$ edges (i.e., possible routes connecting the nodes).

The number of options very quickly exceeds the capability of a human planner to evaluate the different options and find the most efficient solution for a certain transportation problem in the network. Problems of this kind are described and solved by operations research, a hybrid scientific domain blended from mathematics and econometrics. The standard OR approach for transportation (network) problems is linear optimisation or linear programming. The methodology was invented in 1939 by Russian mathematician Leonid Kantorovich, who solved a complex production problem with it, and finally received the Nobel Prize for his invention in 1975. The second problem solved by an LP methodology (by US mathematician Frank L. Hitchcock, in 1940) was already a transport problem.

The term 'programming' does not stand for computer coding in this context, but represents mathematical jargon for 'optimising'. Similarly, 'linear' does not mean that the mathematical problem is linear or in any way simple. Instead, the term refers to a core element of the LP methodology: the allocation problem is solved by a *linear* target function, combining a target value (e.g., maximum profit, minimum distance or minimum time) with a number of relevant constraints, represented by *linear* inequations. In a network problem, constraints could be the limited capacity of vehicles, limited speed or the limited match between demand and the offered service.

The transformation of the original allocation problem into a target function with constraints also leads to a geometrical interpretation of the LP methodology: the target function is represented by a multidimensional space (in its most basic manifestation: a two-dimensional plane), limited by lines. If the problem is solvable, the inequation lines (representing the constraints) separate an interior (the solution space of the target function) from the exterior (non-solutions).

The graphic result for a two-dimensional solution space is a convex polygonal shape (respectively, a polyhedron for more than two dimensions). The optimum of the target function is located somewhere on the outline of the solution polygon, for instance in the corner with the highest value relative to the respective direction of optimisation. The LP algorithms use this fact by ignoring the interior of the solution space, and just moving along the outline until the maximum value of the target function is found. This procedure is illustrated in Figure A.1.

There are a large number of viable LP algorithms, differing by type of possible result (integer versus non-integer) and speed of finding the solution. Standard LP solutions feature polynomic speed, which is safe, but not quick enough for certain problems. Accordingly, George Dantzig, a US mathematician, developed the simplex algorithm in 1947.

The simplex approach could, in the worst case, have an exponential runtime, but in practice it is mostly quicker than traditional LP solvers. Its major upside, however, is its special set of rules for defining the outline of the solution polygon, and searching it for the optimal solution. If the allocation problem to be solved occurs very often and with only slight changes, this approach saves a lot of time, since every new solution can be based on the previous ones ('warm start' instead of 'cold start'), as the rules ensure a learning effect.

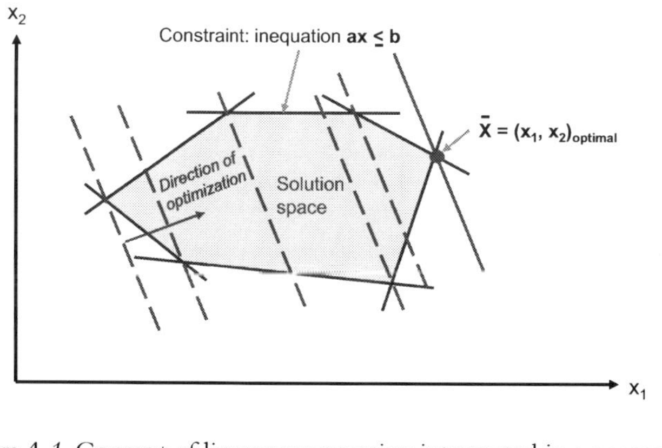

Figure A.1 Concept of linear programming interpreted in a geometrical way

A specific subset of linear programming algorithms deals with optimisation problems featuring at least one variable that is not continuous but discrete. For a transportation problem, this could, for instance, be the number of vehicles deployed on a certain route, since an airline can decide to use one aircraft or two, but no fraction in between. In mathematical terms, these variables belong to the sphere of *integer numbers* (0, 1, 2, 3, 4...), which are a subset of *rational numbers* (i.e., fractions of two integer numbers: 1/1, 1/2, 1/3, 1/4, ...; 2/1, 2/2; 2/3, 2/4, ...; 3/1, 3/2, 3/3, 3/4, ...) as well as of *real numbers* (i.e., all rational numbers plus all non-periodic decimal fractions, such as 1.9730185...).

An 'integer number' problem sounds like a less complex case than a 'real number' problem, since there are infinitely more real than integer numbers (both spheres are infinite, however: real numbers only represent a higher degree of infinity than integer numbers). For linear programming, though, the opposite is true: an integer (linear) programming task (IP) is much more complicated than a (real) linear programming task (LP), since the n-dimensional convex polyhedron illustrated in Figure 6.1 for $n = 2$ as a polygon is not clearly defined; the constraining inequations separating the solution space from the non-solutions outside the polyhedron are not lines in this case but a set of discrete points (those points on the lines with integer values).

The same is true for the solution space itself: if only integer values are acceptable for the target function, the solution plain (for $n = 2$) degenerates into an array of points. Thus, it is much more difficult and time-consuming for the IP algorithm to walk around the (not clearly defined) polyhedron and find the optimum value.

In real-life allocation problems such as in transportation, there is often a mix of integer and real number variables (e.g., the number of aircraft allocated to a certain route can only be an integer number, while price or time are continuous variables). This hybrid optimisation problem is consequently tackled by hybrid algorithms: the so-called mixed integer linear programming solvers (MILP).

Logit models for the description of buying criteria under competition

In the unstable or moving scenarios described above, an initially and intrinsically *efficient allocation* of resources (for a single airline, especially for a monopolist) can very quickly turn into an *inefficient allocation* as soon as competitors change their offerings. The initial airline may decide to react to the move of its competitor(s), leading to an even fiercer competitive scenario, or to a new equilibrium (see further details in the case study in Chapter 8). In any event, the actions and reactions of competitors, especially those described by discrete values, cannot be translated into linear constraint functions.

For this complication of the transportation problem, mathematicians and OR experts have invented the so-called logit model approach. A logit model is designed to perform regression analyses for multiple interdependent *discrete* variables. Similarly to the phenomenon described above (LP versus IP algorithms), a logistical regression (for discrete variables) is much more complicated than a linear regression (for continuous variables). A linear regression model could indeed predict the probability that a variable assumes a certain, discrete value (e.g., 0 or 1 in the binary case), but it may predict a value outside the relevant (integer) range – a result that would be invalid.

A logistical regression avoids this problem by transforming the expectation value of the respective (discrete) variable into a format that is, for sure, discrete itself. The 'logit' serves as a linkage function between the linear regression (with potential non-discrete results) and the discrete probability value (0 or 1 in the binary case). A logit function ηi has the following characteristic logarithmical format:

$$\eta i = \text{logit}\ (Yi=1 | xi1,\ldots,xik) = \ln\ [\pi i\ /\ (1-\pi i)] = \beta 0 + \beta 1 \cdot xi1 + \ldots + \beta k \cdot xik$$

In the case of the logistical regression, the probability πi that a certain event Y assumes the value 0 or 1 does *not* directly depend on the describing variables xi (as with the case of the linear regression) but only *indirectly* via the logit ηi. Likewise, the impact of the calibrated coefficients βk on the probability πi cannot be directly evaluated. The typical format of the transformed 'logistical distribution function' or 'response function' $F(\eta_i)$ looks like the following equation:

$$F(\eta) = \exp(\eta)\ /\ (1 + \exp(\eta)) = \pi i$$

The application of the logit approach for our transportation/resource allocation problem *under competition* is the following: the first planning step for a network is usually the forecast of potential passengers on all relevant routes/O&Ds (i.e., expected demand flows). An airline that is the monopolist on a certain route will attract the entire demand for this air route unless it chooses a ticket price that deters potential passengers. As soon as a second or third carrier enters this market, the clients have a choice, and they will check various parameters to make their decision. The decision factors are usually a mix of hard factors (availability of certain O&Ds via that route, total duration of flight(s) on that O&D, non-stop versus connected service, number of frequencies offered per day/week, timing of the frequencies, product features included, etc.) and soft factors (perceived service level, reputation of carrier, bonus programme, involved partners, etc.).

Every single client could of course feature his individual decision behaviour based on these decision factors, but a large number of potential clients usually show typical tendencies with respect to their preferences for the above-mentioned hard and soft factors. If that is not sufficiently the case, the remedy could be to divide all clients into segments with more heterogeneous decision behaviour – e.g., business and leisure travellers.

In any event, the hard and soft decision factors mentioned above can be used as the describing variables xi in the logit model, since they ultimately trigger the decision of clients for a certain carrier on a certain route or O&D. The coefficients βk can be used to adjust the expected or measured weight of each describing variable in the overall response function. For example, business travellers are usually more sensitive to the timing and number of frequencies offered than to service levels or airline reputation. As stated above, the impact of the coefficients on the final output (the probability that clients buy a ticket from a certain carrier) cannot be directly calculated in the logistical regression case, so calibrating these coefficients is a lengthy process that requires a lot of experience and understanding of market mechanisms.

Appendix B: Big data use case example

The following IoT use case was realised with ParStream's big data technology in 2016.

The Chinese authorities, as well as China's utility companies, have recently assigned a high priority to the generation of sustainable and renewable energy. Thus, they started around 2010 to build enormous wind parks, with tens of thousands of huge wind generators, in the South China Sea. Every single turbine is equipped with adjustable blades, a rotatable head, sensors and a central processor capable of sending hundreds of state variables per second to an onshore control centre. In the beginning the operating companies planned to adjust the position of the rotor heads and blades once in a while, whenever the sensors indicated a change in the wind direction or force – more or less equally *for all* wind generators.

Unfortunately, the first test phases indicated that this assumption was far too simplistic. In the real world, the extraction of wind energy in these wind parks was so massive that the wind turbines significantly influenced the speed, temperature, turbulence and local direction of the wind – and they also influenced each other. The result of this misjudgement was disappointingly poor efficiency on the part of the wind parks; in the initial setting, it was a mere coincidence if single rotor heads had the ideal orientation and blade position at a certain point in time.

Consequently, the steering concept had to be adapted, establishing big data technology. The goal was now to adjust the orientation of rotor heads and blades not every hour or every minute, but *every second* if necessary – individually, for every single wind generator! A much more complex fluid dynamics model was designed to accommodate the interaction between the wind turbines and the wind, as well as between turbine and turbine. From the right perspective, the movements of the turbines now resemble those of huge swarms of birds or fish, executing amazingly elegant manoeuvres, following an invisible swarm intelligence. The result of this upgrade was a genuine real-time IoT use case, with billions of bits of information to be processed every second in order to maximise the efficiency, achieving an increase in the energy output of almost 15 per cent – worth tens of millions of US dollars per year.

As a side note, the data flows between the wind generators and the onshore control centre increased so massively in the course of the upgrade that the initially envisaged *cloud computing* approach failed to deliver the required performance. Instead, a new hybrid approach, called '*fog computing*', had to be developed: the central control unit delegates parts of the complex calculations back to the processors of the individual wind turbines, so that the central and peripheral units alike now share the enormous calculative burden of the real-time IoT concept.

Index

Note: Page numbers in *italics* refer to figures; numbers in **bold** refer to tables.

Printed in Great Britain
by Amazon

56017868R00113